REMINISCING ON BASICS

ABOUT THE AUTHOR

 JOHN MUCAI HOLDS A PH.D. in Business Administration from the University of Nairobi. He is a Certified Public Accountant of Kenya too. He is an alumnus of United States International University, where he graduated with an MSc in Management and Organizational Development and cum laude in BSc in Information Systems & Technology. He retired from Coca-Cola East & Central Africa Ltd in 2017 and has since been pursuing various hobbies and entrepreneurial interests.

REMINISCING ON BASICS

Fascinating Science and Maths Ideas for Everyone

John Mucai

To my wife, Susan, my son Allan, and my daughter Anne, who are the primary source of inspiration and encouragement for the MUCAI Quick Read Series.

For any further information, contact John Muigai Mucai at the following address:
P.O. Box 2069 - 00606, Nairobi, Kenya. Email: johnmucai@gmail.com

Cover design by Linda Matama

ISBN 978-9914-702-90-3

"The more you know, the more you know you don't know."

—ARISTOTLE

CONTENTS

LIST OF TABLES

LIST OF FIGURES

FOREWORD

Most MUCAI QUICK Read series books reflect on past events and lessons learned from the experiences. The current book is not different. However, in this instance, the focus is on a few interesting, and in some cases, incredibly fascinating science and mathematics concepts encountered by the author.

For scientists and mathematicians, the book may perhaps be overly simplistic. But hopefully, the anecdotes in the book will provide some amelioration.

Other mortals will hopefully find the science and mathematics concepts interesting. They will most likely realize that they have missed out on fascinating knowledge. And hopefully, that realization will motivate them to delve deeper into the fantastic vistas of knowledge that exist out there.

The material in the book is presented in a manner that is easy to understand.

The chances are that many readers will have already encountered many of the ideas in the book in their early high school days. However, the author attempts to shed light on the exciting aspects that may have been glossed over or omitted in high school.

The author's primary objective is to make many readers people more inquisitive about science and mathematics, irrespective of their age or current station in life.

The reader should no longer accept anything as a given, even what might seem like a hard mathematical or scientific fact.

PREFACE

The epitaph at the beginning of this book encapsulates this book's essence, namely, that "The more you know, the more you realize you don't know." This is true in almost any sphere of life, and particularly, and perhaps surprisingly, in science and mathematics.

As I have been reflecting on some of these things, I thought that in the spirit of the MUCAI QUICK Read series, it would be good to share some of my sense of wonder with readers.

Hopefully, the things I present in the book will serve as metaphors for how we need to think about life. No matter how simple or complex some things might seem, there is something about them hidden from us by nature. And therefore, the more that we should respect everything in the physical world around us.

There are many open questions in science and mathematics. There is a lot out there that awaits discovery. Teachers of science and mathematics should continually use such knowledge gaps to stimulate intellectual interest in their students.

Science and mathematics are full of wonders. And one does not need to be a student, scientist, or mathematician to partake of the surprises.

I have written the book for ordinary world citizens who are inquisitive about nature in all its real and abstract forms. Indeed, the reader does not need to have any specialized knowledge of mathematics or science to appreciate the book's ideas. Common sense is enough. But clearly, basic science and geometry learned in junior high school will be an advantage.

I hope that many readers will find some of the ideas presented in the book exhilarating – the kind of exhilaration that occurs after partaking in a good piece of new music, art, or poetry. And if this happens, then one of the objectives of the MUCAI Quick Read series would have been accomplished.

I hope the reader will be inspired to pursue the ideas in more detail and possibly uncover some new mysteries of nature.

ACKNOWLEDGMENTS

The Almighty God has been the shining guiding light throughout my life, even in this book project. I will always remain steadfastly thankful to Him.

This book would not have been possible without the ongoing unshakeable support of my wife, Susan, my son Allan, and my daughter Anne. I am deeply grateful to them.

I would also like to thank Teddy Muhia and Nzisa Kattambo for editing the book.

INTRODUCTION

When I was in junior high school, I came across some ideas that I thought were fascinating. But I never got a good chance to delve deeply into them. I was more focused on passing exams.

The exams came and went. And I continued with other things in life. All those fantastic ideas receded into my memory, especially as I pursued the things that would secure a decent livelihood for me. In this particular case, a career in the world of business.

This book is for those, like me, who never lost the appeal for maths and sciences, but had to move in different directions in life for various reasons.

Recently, I thought it would be interesting to revisit some of those ideas to find out whether I could find newer, more fascinating insights. And, more importantly, to examine the relevance of those ideas in contemporary terms. I was not disappointed.

As I started taking a closer look at the ideas, I was humbled in a refreshingly new way by the enormity of the intellect of people like Pythagoras of Samos (c. 570 – c. 495 BC), Aristotle (c. 384–322 BC), and Euclid (c.365 – 300BC); Sir Isaac Newton (1642-1727), Gottfried Wilhelm Leibniz (1646-1716), and Leonhard Euler (1707-1803); and many others. These few individuals built the foundations of science and mathematics in a truly remarkable way. And in many instances, literally from a clean slate. The world owes a lot to these intellectual giants.

Perhaps we can learn a lot ourselves, irrespective of our education and careers, by going back to the basics and gaining a more comprehensive understanding of the fundamental ideas that are the cornerstone of contemporary knowledge in different domains.

For example, when I studied high school calculus, I vividly remember how amazed I became when I grasped the idea of limits. What truly amazed me was its simplicity and how that simplicity had been taken to the "limits" by mathematicians and physicists to solve some very complicated problems. I still find the idea quite baffling. I revisit this central idea of calculus and suggest that other aspects of human life are awaiting major discoveries by applying similar ideas.

Everyone who has gone to primary school knows the meaning of gravity. The discovery and proper documentation of this invisible force are attributable to Isaac Newton. He published his theory of gravity in 1687, a mere 333 years ago. There is evidence that people understood it? But the more I think about it, the more I feel that I do not understand it.

Another fascinating concept is latent heat. It is an interesting subject that I did not get enough of when I was in high school. The subject continues to intrigue me up to this day.

The subject of tropisms made a lot of sense those many years ago in biology class. But today, it no longer seems to make sense to me. It is wrought with many unanswered questions.

And then there was the theory of ecosystems. This was the final topic in my biology class in high school. I could not get enough of it, either. I was a little disappointed that it came right at the end of the course. I wished the teacher had covered the subject in greater detail. I discuss the basics of this subject and hope the reader will share in my sense of awe in some of its fascinating aspects.

And then there is chaos theory. When I first encountered this topic, I was shocked. My sense was that Benoit Mandelbrot, and the others who discovered it, had suddenly found the secret of life. An alternative to the theory of evolution. Something more powerful than Darwin's theory of evolution. I sincerely believed that it would be the next big thing in science. But that did not happen, and I am still puzzled why that is so. There are many other fantastic ideas presented in the book.

One final point. Please do not be put off by the few basic formulas in the book. They are included in strategic places within the text simply to keep the mathematicians happy. Feel free to shut your eyes when you encounter the formulas.

PART 1

Science Anecdotes

CHAPTER 1

Gravity

*"Everything every day here on earth is based on gravity, and
you don't realize it until you don't have it anymore."*
— Peggy Whitson

WHAT IS GRAVITY? THE ANSWER to this question
may depend on where you come from and your
upbringing. Such a word does not exist in my local
language.

If you had asked my grandmother this question, she
would probably have told you that it was a word invented by
the "wazungu" (Europeans) to confuse Africans.

If you were to ask a lawyer the same question, they would
probably respond without blinking that gravity is a term used
by judges. Further, judges use the term to express the
seriousness of an offense, giving them latitude to mete out
the appropriate level of punishment to an offender. In other
words, whatever decision the judge makes reflects the
gravity of the matter. But exercise your discretion wisely

before you ask a lawyer such a question, as you could potentially receive an invoice that would give a whole new meaning to the gravity of your question.

If you asked the question to a trapeze artist who has sustained an injury during a circus performance, you would get an answer very close to the scientists' idea of gravity.

But perhaps the best answer you might get would be from a ten-year-old whose ice cream has fallen to the ground after one sip with his tongue.

Gravity Anecdote

An interesting incident happened in Kenya on March 27, 2016, that put a new spiritual, albeit comical, meaning to the idea of gravity.

A prominent political party leader in Kenya was addressing a large crowd of followers at the Karisa Maitha Stadium in Malindi.

The party leader had just finished narrating the story of how he had been taken to the mountain in Jericho, the same mountain where Satan had tempted Jesus. He said that Satan had shown Jesus the vast earthly Kingdom he could inherit if he decided to prostrate and worship Satan. The politician said that Jesus declined the offer and told Satan: "*nenda mbali shetani*" (go away, you devil).

Everyone in the stadium burst into uncontrollable laughter. Then something else happened that was completely uncontrollable.

In the midst of the laughter, the dais on which the party leader and several of his supporters stood collapsed. Luckily, nobody sustained a serious injury.[1]

The incident attracted a lot of interesting feedback on social media. Someone said it was like a flash of lightning from God as a warning for messing around with the gospel for political reasons.

But scientists can say with certainty that the party leader and the people standing on the dais experienced a generous dose of the force of gravity.

What is Gravity?

Surprisingly, something as simple and as familiar as gravity was not formerly discovered until 1687, when Sir Isaac Newton published a mathematical theory on the subject. One wonders how, prior to that discovery, scientists interpreted the simple phenomenon of things falling to the ground.

It may be reasonable to assume that scientists did indeed believe that there was a force that tended to push things downwards, but it was Newton who put some structure around it.

But even today, there is a subtle aspect of gravity that may not be clearly evident to many people.

Gravity is not merely the force that pulls things to the ground. It is the force that attracts objects to each other. This is different from the physical attraction between individuals, loosely termed "chemistry."

Therefore, one wonders what other phenomenon exists today that people just take for granted that another Newton will "discover" and describe in a way that will surprise everyone.

* * *

So, what is gravity, really? One may ask.

Newton determined that gravity accounted for small objects on earth and large objects in the universe. Newton's basic theory was that big things tended to attract smaller objects. For example, the earth draws the moon to itself, but the moon does not come crashing down on the earth because of other countervailing forces.

Newton also discovered that the amount of gravitational force was correlated to the mass and distance between objects. His theory was that for small objects on earth, the gravitational force is equal to the mass of the smaller object multiplied by the object's acceleration. Acceleration in this context is equal to 1 meter per second squared.

$$\vec{F} = m\vec{a}$$

Where:
$\vec{F}$ is the gravitational force
m is the mass of an object
$\vec{a}$ is the acceleration of the object

According to Newton, the further away objects are from one another, the smaller the gravitational force between them. (We will revisit squares and square roots a little later in the book.)

Newton's famous formula was that the gravitational force between two objects was proportional to the masses of the two objects multiplied by each other, divided by the distance between the two objects squared.

$$F = G\frac{m_1 m_2}{r^2}$$

Where:

F is the force

m_1 is the mass of one of the objects

m_2 is the mass of the other object

r is the distance between the centers of the objects

G is the gravitational constant

Newton did not establish how big or small the constant G would be. This was determined more than 100 years later, in 1798, by another mathematician called Henry Cavendish, who did some careful measurements of gravity and came up with the following value of G: 6.67×10^{-11}.

When Newton said that there was an invisible force that caused things to fall down and people opened bottles of champagne to celebrate the discovery, did anyone ask where the so-called force came from? And what that "force" was anyway?

Interestingly, the answer came 228 years later. Albert Einstein came up with the solution. He discovered that the gravitational force was due to what he called space-time curvature. Einstein backed up his idea with sophisticated mathematics, but it is really an easy concept to understand.

The space-time curvature explanation goes along the following lines: think for a moment of space as a large flexible sheet. If you place a big object in the middle of the sheet, such as a football, the sheet will cave inwards due to the ball's weight. And if you place another smaller object on the sheet of cloth, it will roll downwards towards the big ball due to the curvature created by the big ball. Figure 1 shows an example of this phenomenon.

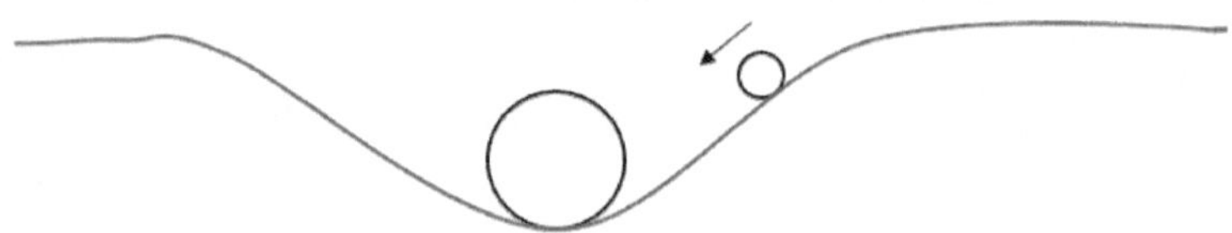

Figure 1: Einstein's Space-time curvature

Einstein's explanation was described as the theory of general relativity. According to this theory, the gravitational force is the rolling motion that occurs as a smaller object rolls towards a bigger object in the space-time curvature depicted in Figure 1. Scientists describe it as the warping of space-time. [2]

There is a lot of math and science behind this explanation beyond the scope of this book.

From a layman's perspective, the challenge of the explanation is that it makes sense when one thinks of space as a flexible sheet. This is clearly not the complete picture since space is not unidimensional, like the flexible sheet. We can only trust that our scientists have examined the theory from a multi-dimensional angle and are happy with Einstein's explanation.

But we live in a different world today where formal research in almost any conceivable phenomenon is in progress in different parts of the world. So, something as earth-shattering as the discovery of gravity seems inconceivable. Or is it? How about the passage of time? What is time anyway? Could it be a force, too, such as gravity - a force that cannot be stopped?

Theorizing is Fun

Let us have a little fun with some theorizing. We all have a shared experience of time as a combination of past, present, and future. But let us think a little more deeply about this idea.

If I set the stopwatch to run for five minutes, I can easily tell when the five minutes end. But what are the five minutes, really? How about one minute? How about one second? What happens between the first and the second "second"? What is contained in the space between these two points? If we could understand what exists in that tiny space, perhaps we could understand what time really means. If we could slice that space into tinier and tinier segments, we would clearly end up with nothingness.

Just contrast this with material objects. Suppose I took the sheet of paper lying on my desk and divided it into smaller and smaller pieces. In that case, I would reach a point where there would be nothing to split anymore. Scientists will say that physical matter as we know it will cease to exist at that low level. It will change to a quark, the smallest form of matter known to man.

But what would happen if we split the quark into even smaller pieces? Then we would likely reach a point where there would be nothing else to split. The non-quark would be the same as the nothingness we saw when dividing time into smaller segments.

There would be no difference between time and quarks. Accordingly, one could conjecture that physical objects made of multiple quarks are, in essence, made of numerous time segments. Physical objects are manifestations of time segments. We would be justified in arguing that nothingness can manifest itself as time segments or quarks of physical matter at that low level of nothingness.

The same principle applies to space. Let us consider space as the distance between two objects, such as the two footballs shown in Figure 2.

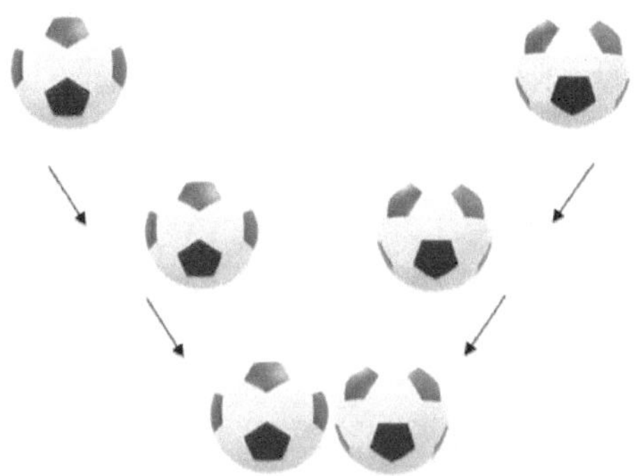

Figure 2: Bringing two footballs closer to each other

If we bring the two footballs closer and closer to each other, the space between them continues to shrink. In fact, if we keep bringing them closer in small minute steps, there is one last infinitesimally small step that will remain before the balls touch each other. If we could measure this little tiny bit of space, it would be equal to nothingness itself. We encountered the same nothingness when we tried to split time into small segments. It is the same as the nothingness we met when we attempted to divide a quark into smaller bits of matter.

So, you can pour yourself a glass or mug of your favorite drink as you have just discovered a new universal theory of nothingness. Let us give it a fancy name, the Mucai Grand Unified Theory of Infinitesimals (MGUTI). It is quite some fun, isn't it?

But comedy aside, this is the kind of thing that makes science and mathematics great fun. The freedom to travel to the edges of knowledge where pure bliss exists. A place where one can exercise their intellect in new ways.

And before you finish laughing off the idea of infinitesimals, let me mention that this idea is at the heart of the theory of limits. This theory is the essence of the calculus invented by Sir Isaac Newton and Gottfried Wilhelm Leibniz about 350 years ago. This idea revolutionized mathematics and science.

We will explore the theory of limits a little later in the second part of this book. Its simplicity and complexity, all tied together, not to mention its remarkable practical applicability in the physical sciences, makes it a truly remarkable theory.

CHAPTER 2

Tropisms

"Plants do not speak, but their silence is alive with change."
—May Sarton

WE HAVE SEEN THAT THE force of gravity is so strong that it causes prominent politicians to fall through daises. Why do some other things, such as plants, seem to defy gravity and grow upwards? But even more bewildering, why do plants not grow upwards perpetually? They usually reach a limit where they just stop growing upwards. Why is this? And what determines the level at which this upward growth stops?

Nature seems to be full of puzzles. These are the mysteries that scientists have been grappling with for centuries.

Occasionally, scientists are rewarded with fascinating insights into the workings of nature. Sometimes they earn Nobel prizes for sharing their ideas regarding the workings of nature.

What is Phototropism?

I discovered the answer, or let us say, a partial solution to the upward growth in defiance of gravity when I was in primary school.

After a few engaging, practical lessons involving the dissection of frogs, it was time for some experiments using plants. For the first experiment, the teacher put some seeds in a petri-dish. He then placed the petri dish in a box. The box had a small hole near the top edge on the side furthest from the petri dish (similar to the small window in the police cell described in the Tropism anecdote above).

The teacher then sealed the box, except for the small hole, and placed the box on a table in one corner of the room. He said that the purpose of the small hole was to let in air and some light. He did not tell the students what would happen next.

A few days later, the teacher cut one side of the box to reveal the seed in the petri dish. I will never forget my feeling of awe. The seedling shoots had miraculously bent and grown towards the corner of the box where the light was coming from. It was incredible.

(The inmate's actions were similar to those of a plant stored in a sealed box, with only one small source of sunlight.)

I wondered how the plant had figured out where the light was coming from. Did the plant have a brain that prompted it to move towards the sun (like the inmate)? Or was it a magic trick conjured up by the teacher?

The teacher told us that the phenomenon we were witnessing was called phototropism - plants' natural tendency to grow towards the light. He said that light was necessary for the plant to develop chlorophyll, essential for plant growth. Those were the scientific facts.

The teacher explained that tropism was just one of the many external stimuli in the environment that affected plant growth. Other examples included gravity and water.

The effect of gravitational stimuli was called geotropism (also called gravitropism), which caused plants' roots to grow downwards. The growth of plant roots towards the water was called hydrotropism. These were clear and logical explanations of plant growth, but they raised more questions than answers.

One nagging question was: what actually made plants sense the environmental stimuli in the first place?

And sure enough, I found the answer later on in high school.

* * *

Plant growth is controlled by a group of hormones called auxins found at the tips of stems and roots of plants.

Auxins are produced at the tips of stems. They tend to accumulate on the side of the stem that is in the shade. Accordingly, the auxins stimulate growth on the shaded side of the plant. Therefore, the shaded side grows faster than the side facing the sunlight. This phenomenon causes the stem to bend and appear to be growing towards the light.

Auxins have the opposite effect on the roots of plants. Auxins in the tips of roots tend to inhibit growth. So, auxins that accumulate at the tip of the root will impede growth. If a root is horizontal in the soil, the auxins will accumulate on the lower side and inhibit growth. Therefore, the lower side of the root will grow slower than the upper side. This will, in turn, cause the root to bend downwards, with the tip of the root growing downwards – a manifestation of gravitropism.[3]

* * *

There were yet other questions that emerged from these explanations. Firstly, where did auxins come from? Secondly, why did auxins tend to accumulate at the tips of stems and roots? Thirdly, why did auxins have different effects on stems compared to roots?

Answers to these and many other related questions were probably accessible in more advanced levels of biology.

But why all the fuss about tropisms? The answer is that, to non-biologists like me, they represent one of the fascinating aspects of nature. Some things seem simple, and yet they are so complex.

Tropism Anecdote

In 2016, one of my brothers was traveling home on a bus. As he was alighting from the bus at Uthiru shopping center, Nairobi, a pickpocket stole my brother's cell phone and identity card. It was quite an annoying experience.

Operating without a cell phone or identity card was going to be difficult. The first step in securing a temporary official identification document, pending the receipt of a duplicate identification card, was to report the incident to the police.

Luckily, if there can be such a thing as luck in such circumstances, my brother did not have to go far. The Kabete Police Station was just a few minutes' walk from the bus stop.

* * *

As my brother was walking away from the police station after filing his report, he heard somebody calling him from a back window of one of the offices adjacent to the office where he had just reported the theft.

The sound stimuli from a stranger prompted my brother to turn back. He walked towards the window to talk to the person calling him. But little did my brother know what he was getting himself into.

It turned out that the window from where the caller had whistled to my brother was a police cell. The small window was the only source of sunlight for the inmates locked in the cell. An inmate had stretched his head to the window through which he had seen and called my brother.

As my brother moved towards the caller, he did not notice a policeman approaching from the opposite direction.

When my brother edged closer to the window, he heard a loud voice behind him: *"simama hapo, mara moja!"* (Stop right there, immediately!). The command was from the policeman my brother had seen a minute earlier.

My brother was arrested on the spot. The charge: attempting to pass contraband to an inmate.

My brother was dumbfounded. What contraband was the policeman talking about?

* * *

It turned out that inmates had occasionally been found with cigarettes. The source of the cigarettes had remained a mystery until that day when the policeman thought he had "caught the culprit red-handed."

My brother was frog-matched back to the police station and locked up. The events that followed after that are reserved for his future memoirs. For now, suffice it to say that it took another six months for my brother to clear himself from a false charge. It was the sort of stuff that fairy tales are made.

In this example, we can see the effect of "phototropism" on a cell inmate that almost caused an innocent person's incarceration.

The real name of this type of behavior is "cantankerousness." The behavior is associated with people with a "peeping tom syndrome" history.

Tropisms and Consciousness

Turning back to plants, a basic understanding of tropisms was exciting enough to sustain my curiosity in the sciences. This was until I changed course and moved to other more attractive avenues in terms of career prospects.

* * *

My curiosity was triggered considerably when I recently realized that science had progressed way beyond the basic ideas of tropisms into questions of whether plants possess consciousness. The ideas around this subject are well captured in the work of the Italian neurobiologist Stefano Mancuso, Professor at the University of Florence.[4,5]

According to Mancuso (2018), although plants may not have organs like animals, they nevertheless can perform certain functions that require consciousness, just like animals. For example, plants can be trained to memorize, solve problems, communicate, and even socialize.

Mancuso has performed experiments to demonstrate that plants can keep a memory for two months, unlike insects that have a two-day memory. He has also shown that plants coalesce into social networks through networks of roots. The networks facilitate the sharing of information and even nutrients. Through the root networks, plants can warn each other of danger through the emission of chemical distress signals. Plants can also detect sound.

Mancuso considers consciousness as the same thing as self–awareness. And to demonstrate that plants are self–aware, he describes how plants that grow under the shade of another tree can accelerate growth to reach the light above the shade. Interestingly, shoots of a plant shaded by other leaves of the same plant do not exhibit similar competitive behavior. They seem to be aware that they are part of the same plant.

Mancuso further points out that numerous experiments have shown that plants' roots are sensitive to about 20 different chemical and physical parameters. Humans cannot sense some of these parameters.

While animals have organs for specific functions, plants evolved differently. They developed a decentralized system whereby the entire plant performs all functions simultaneously.

Some of these ideas raise some interesting philosophical questions.

CHAPTER 3

Phase Transitions

"When I feel the heat, I see the light."
— Everett Dirksen

WE CONCLUDED THE FIRST CHAPTER by looking at what happens when we divide things into smaller and smaller bits. Now let us look at what happens when matter transforms into different states – what physicists call phase transitions. And there are many kinds of phase transitions.

Phase Transitions Anecdote

One December in the early 1970s, an unusual event occurred at Kinoo. A truck and trailer carrying hundreds of crates were involved in an accident. The trailer overturned, and the door swung open, spewing the contents of the trailer on the roadside.

My brothers and I, along with some neighbors, heard a loud bang as the overturned trailer hit the ground, followed by the sound of crashing bottles. We knew right away that a motor vehicle accident had occurred and immediately rushed to the scene.

What we saw when we arrived at the scene was beyond belief. There were bottles of beer scattered all over the place. There were also multiple crates of beer, some still full of beer bottles.

But what was even more surprising was that the passersby who witnessed the event had spontaneously decided to help themselves to the largesse of beer.

Now lying on its side, the trailer had become a beehive of activity. There were tens of people scrambling to remove as many beer bottles and crates as possible from it.

Some people had even started drinking the beer as they stood on the sidelines, watching the frenzy, as in a rugby match.

The driver of the truck was totally helpless. He just watched in horror as his cargo was looted. During those days, there were no cell phones, so he could not call the police for quick help.

After a short while, even vehicles passing by stopped to participate in the looting. Everyone just seemed to have gone berserk. The trailer was emptied in less than an hour, by which time it had started getting dark.

News of the free booze spread around Kinoo village like wildfire as people were seen carrying crates of beer on their shoulders or several bottles held tightly on their chests.

And then, something even more unusual happened. As the crowd grew larger, those who arrived after the trailer had been emptied became agitated. The newcomers started surrounding the main and substantially larger truck. The main truck was still in good condition. It did not appear to have been seriously damaged. But before we knew it, people started chanting what sounded like war cries, but full of freakish laughter, as they started banging the truck's door locks. The bolts flew away like bullets, and the truck door swung wide open. People immediately began removing the contents in a mad frenzy.

The orgy of theft and the attendant confusion were mind-boggling. If the earlier bevy of activity around the fallen trailer was turmoil, the main truck's looting was utter chaos. The tens of people scrambling to extract beer from the truck looked like a million Serengeti gazelles, wildebeest, and zebras crossing the Mara River in the Great Migration.

Interestingly, some people who had taken crates of beer from the fallen trailer returned for a second serving.

That night, the per capita consumption of beer in Kinoo village must have hit historic levels. And by an incredible coincidence, it was smack in the middle of the merry-making season. It was Boxing Day. It was as if Father Christmas had come to Kinoo one day late.

The events that played out in my father's homestead later that evening were unprecedented. Although many details have faded from memory, such as the names of some of the characters involved, the broad outline of the events as they unfolded is still fresh in my mind.

Alphonse (not his real name), a migrant from Tanzania, employed by my father as a casual laborer in the family *shamba* (small field), had secured about four crates of beer. He was a teetotaler, but he had landed on a fortune of free beer on that day, and he was not going to let it go to waste.

During the first one hour after he returned home with the beer, he was full of himself. He narrated how he had penetrated the thick mob of people to retrieve the crates of beer.

The narration was full of jokes. We laughed our lungs out as Alphonse recounted the intricate details of how he had skillfully outwitted others in the crowd.

As the laughter subsided, it suddenly dawned on him that something had to be done with the beer. He quickly obtained a bottle opener and started opening the bottles and sharing them with those around, including teenagers who had never sipped even a drop of beer in their innocent lives. Alphonse helped himself to the beer too.

I can still see in my mind's eye Alphonse taking a big gulp of beer with enthusiasm. After the gulp, he sucked his cheeks inwards. He closed his eyes firmly, evidently in involuntary disgust after experiencing the beer's sharp bitterness. He did not open his eyes for several seconds as if to let the pain in his mouth and throat settle.

After the initial shock of his sensory organs, Alphonse managed to control himself. He consumed the remaining contents of the bottle at a slower pace. Evidently, he was forcing himself to drink the expensive bitter liquid, not because of its pleasant taste but sheer vanity. After all, the beer was free and abundant in supply.

But something interesting started happening shortly after that.

* * *

An hour later, a phase transition occurred. Alphonse suddenly became exceptionally joyful. He started talking in a drawl. He also became utterly uninhibited in what he was saying. Several people around him also became incredibly merry. Even the shy ones started opening up and talking nonsense.

We usually spoke in Kiswahili in Alphonse's presence because he could not understand our mother tongue, Kikuyu. But during that phase of the event, some people started talking in English – to impress the others. The loud voices rented the air with broken English, primarily spoken with an American accent. And this ought to raise an interesting psychology research question: Why do some people in Kinoo tend to switch from local languages to English when they become inebriated? This phenomenon is prevalent even today.

* * *

Another hour or so later, there was another phase transition. Some people who were drinking beer for the first time became utterly drunk. Some of them could hardly talk. They were blubbering in a predominantly English drawl. That Boxing Day evening, the English language was used and abused or used to abuse. And most poignantly, common decency was thrown out of the window. Typically unspoken words were no longer taboo.

* * *

After another two hours or so, the crowd dispersed. I cannot clearly remember how the rest of the evening unfolded.

All I can recall is that several individuals woke up the following morning with bloodshot eyes. Some individuals could hardly wake up from bed.

Many of the merrymakers entered the fourth phase that they had hardly expected. They complained of severe headaches, a manifestation of a terrible hungover. That phase signaled that some people would never consume alcohol again. They probably could not imagine having to go through another similar experience.

* * *

And in the semi-final phase transition, Alphonse, who was now ill from over-indulgence the previous night, dished out to neighbors all the remaining beer in his possession.

* * *

Then came the very final phase. I try hard to forget it, but it never goes away from memory. This phase kicked in at around 2:00 pm on the day after Boxing Day.

My father had learned of the overturned beer track and the hullaballoo that had ensued. But he had left it as an item of passing interest.

However, on the following day, through a strange twist of events, he decided to carry out a mini-Spanish inquisition. Let us just call it the Kinoo inquisition. He summoned everyone in the homestead for an impromptu meeting. The venue was the steps in front of the main entrance to the house. He sat on a dining chair. The rest of us sat on the steps below him. It was quite a weird setting, perhaps matching the situation's awkwardness.

All those involved in the shenanigans the previous day and night pled guilty, and so did the main culprit, Alphonse. Everyone apologized profusely and committed to never ever do such a thing again.

My father meted out different types of punishment to the offenders to discourage them from such misbehavior ever again.

I have never known how my father discovered the merry-making, but I have a hypothesis. It was most likely suspicion based on the odd behavior of certain characters.

* * *

Soon after the drunken lot went to bed, the party host, Alphonse, who lived in a small wooden house about 10 meters from our house, must have turned into a frog. His snores must have been so loud and in varying octaves that they could be heard in the North Pole. That must have obviously turned off my father considerably, given that his bedroom was not very far away from Alphonse's living quarters.

The following morning, Alphonse must have been late for work. And as my father went searching for him, he must have seen Alphonse still in a state of stupor with his usually snow-white eyes (with a dark black pupil in the middle) now bloodshot.

Alphonse must also have been reeking alcohol as if he had suddenly emerged from one of the brewing vats at Kenya Breweries. And that would have been Alphonse's undoing.

And if that is what happened, then my father must have undoubtedly concluded that Alphonse had not acted alone.

* * *

In summary, there were six distinct phase transitions in the ill-fated Boxing Day phenomenon: (1) sobriety to a wild frenzy, (2) a wild frenzy to intense joy, (3) intense pleasure due to inebriation, (4) inebriation to splitting hungover, (5) splitting hungover to an inquisition, and (6) punishment.

I sometimes prefer to call phase (1) "the onset of chaos" and phase (6) "utter chaos." The rationale for this nomenclature will become more apparent in the last two chapters of this book.

The technical non-scientific name for all the six-phase transitions is "craziness."

Phase Transitions in Physics

A phase transition in physics is a phenomenon that occurs when a material changes from one state to another. For example, by applying heat to a container of ice cubes, the ice cubes gradually change into water. The heating up of the water continues until the water starts boiling.[6] The moment the water begins boiling is called the boiling point. This boiling point is roughly 100 degrees Celsius. But the exact temperature depends on various other factors, such as the air pressure in the surroundings. Still, on average, it works out to 100 degrees Celsius.

The odd thing is that no matter how long you continue applying the heat to the water, its temperature cannot go beyond 100 degrees Celsius. Why is this, one might wonder?

* * *

Scientists have determined that the additional heat energy that one applies goes towards heating up the water molecules (the tiny particles of which water is made). After some time, the molecules get immensely agitated and knock on each other at high speed transforming the liquid into a gas (water vapor). If the gas is heated further, it disintegrates into plasma.

The transitions as the ice cubes melt into water and as water evaporates into water vapor are depicted graphically in Figure 3.

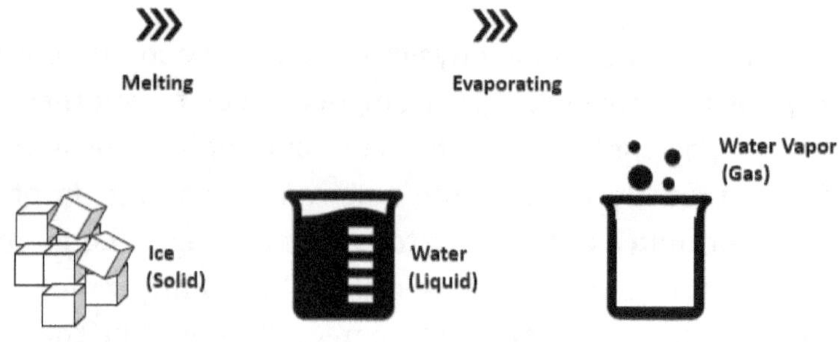

Figure 3: Melting and evaporation

The processes depicted in Figure 3 require energy input to weaken bonds between the components of matter (molecules). Accordingly, the processes are referred to as endothermic.

The reverse processes are shown in Figure 4.

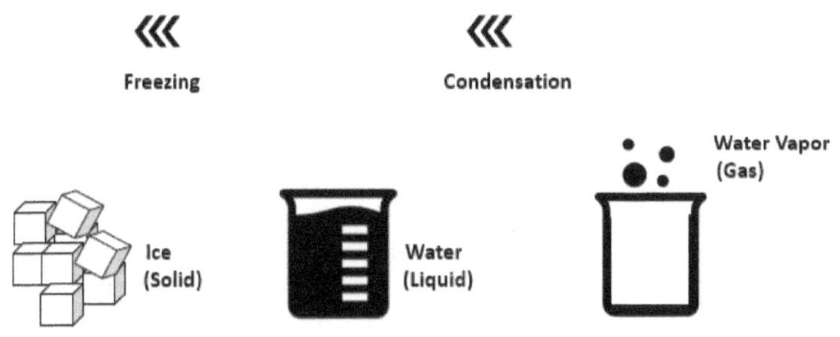

Figure 4: Condensation and freezing

The processes depicted in Figure 4 are exothermic because they involve the absorption of energy during the formation of bonds in matter (molecules).[7]

There are instances when ice transitions directly from solid to water vapor (gas). This transition is referred to as sublimation. The reverse, when a gas turns into ice, such as frosting in winter, is called deposition.

For present purposes, we will ignore the phenomenon where the gas turns into plasma after heating at high temperatures and breaks down into smaller particles of matter (electrons), a process called ionization. The reverse process is called deionization.

* * *

The heat energy applied to convert ice cubes to water is called the specific latent heat of fusion. And the heat energy required to turn water into water vapor is called the specific latent heat of vaporization.[8] See Figure 5.

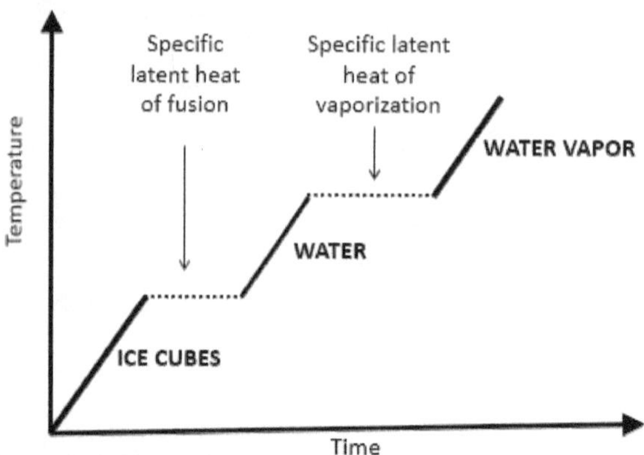

Figure 5: Specific latent heat

Interestingly, the energy required to change water into water vapor is approximately seven times the energy needed to convert ice cubes into water.[9] Why is this?

* * *

Another interesting point to note is that something will be happening at high energy during these transitions without us being aware of what is going on.

Let's take the transition from water to gas, for example. There is a period during which we will continuously be applying a very high temperature on the water without anything seeming to happen. Then, suddenly, the water will gradually start turning into vapor.

* * *

The ideas around specific latent heat are part of the physics of phase transitions. Phase transitions occur at the boundary where matter moves from one state to another—for example, the transition of metal from unmagnetized to magnetized. There are many other types of phase transitions.

To use the words of Gleick (1988) regarding the behavior of matter near the point where it changes from one state to another:

As singular boundaries between two realms of existence, phase transitions tend to be highly nonlinear in their mathematics. The smooth and predictable behavior of matter in any one phase tends to be little help in understanding the transitions. A pot of water on the stove heats up in a regular way until it reaches the boiling point. But then the change in temperature pauses while something quite interesting happens at the molecular interface between liquid and gas.[10]

The question that still lingers in my mind is why phase transitions occur at particular points in the different states of matter? Is there some kind of universal rule book hidden from us, which indicates when certain things in nature should happen, such as when water should start freezing or boiling? If such a book exists, what else does it contain other than the laws of chemistry and physics known to us?

CHAPTER 4

Conservation of Energy

"The law of conservation of energy tells us we can't get something for nothing, but we refuse to believe it."
— Isaac Asimov

NOWADAYS, THERE IS A lot of talk about the conservation of energy. But what is energy, and how can we conserve it? Most of what I read in newspapers and see on television regarding energy conservation is related to conserving non-renewable resources. But turning to the idea of energy, what does it really mean? An anecdote will help set the stage for a discussion of this subject.

Energy Conservation Anecdote

I feel great nostalgia when I remember the types of pranks we played as kids. One particular incident reminds me vividly of the subject of conservation of energy.

Close to our home at Centre in Nakuru was a big playground where kids spent most of their time playing different games. The field was used as a parking lot for upcountry buses in the late afternoon.

At that young age, kids did not know or question why the buses were parked there. But even if an explanation had been given, the kids would not have been overly concerned about it. On the contrary, there was one secret reason kids liked the buses: the unusual enjoyment that the buses provided.

During those days, most upcountry buses had a luggage carrier at the top, with a ladder at the back of the bus used for climbing to store or remove luggage from the carrier. The ladder was to become an enormous source of enjoyment.

Buses would usually enter the field from one end. The entry point was a little rough. Therefore, the buses were forced to slow down.

During the few seconds when a bus slowed down, we would climb on the ladder at the back. We would enjoy the bus's thrilling motion for the next five or so minutes as it drove to the parking spot. That brief ride hanging on the ladder was immensely enjoyable. And even more thrilling was the kinetic energy that we experienced as we lept from the ladder just before the bus came to a standstill.

It was an incredible stunt that created a tremendous adrenalin rush. Also, we had to time the jump from the ladder perfectly to avoid zooming off at high speed and getting injured. We could not wait for the bus to come to a complete standstill for fear of being apprehended by the bus driver or the conductor.

In hindsight, it was a hazardous activity. A wrong move while descending from the ladder could have resulted in a severe injury, especially if one leaped from the bus when the bus was moving fast. But these were dangers that we could not see. The thrill was just too good to warrant any worries about potential hazards.

On a few occasions, an angry driver or conductor, fed up with the nuisance, would run after the kids after the bus had stopped. But in most cases, the kids would be too fast for them. On a few occasions, one unlucky kid would be apprehended and receive instant punishment.

I was among the lucky ones and never got caught – except on one unfortunate afternoon.

* * *

One of the conductors had decided to take exceptional measures to curtail the menace once and for all. He had positioned himself close to the door of the bus. He quickly jumped out before the bus came to a standstill and nabbed Kanja and me just after we jumped off the ladder.

We struggled to wiggle ourselves from his firm grip to no avail. But almost by some magic, Kanja wriggled himself free and escaped at exceptionally high speed. The conductor could not go after him, as it would have necessitated releasing me. He was determined to keep me prisoner so that he could mete out punishment that would serve as a lesson to all the other kids.

I remember crying loudly and asking the gentleman to release me, but he would not yield.

After a few minutes, a small crowd gathered around to find out what was happening. I was full of fear and trepidation as I did not know what the conductor and the bus driver intended to do to me.

The conductor and driver explained to the crowd how kids in the neighborhood had become a menace. He went to great lengths to justify the importance of curtailing the shenanigans once and for all.

I remember some voices pleading for mercy, but the two gentlemen were not ready to exercise compassion. They were determined to do something that would teach all the kids a lesson – and I would be the example.

Fortunately, one sane voice in the crowd suggested that the best disciplinary action was to report me to my parents.

I was then asked to give my parents' names and the location of my home.

I knew that if I was taken to my parents, the consequences would have been dire. For a start, I would probably have had to kiss goodbye to playing in the field for at least one week. I would also have received another form of instant punishment. Because of this, there was no way that I was going to let these gentlemen take me to my parents. The problem was that escaping as Kanja had done was entirely out of the question. The conductor had tied my hands with a rope. Therefore, I would not have been able to run away.

There was no way out. But I thought of another innovative idea: I would not reveal the true identity of my parents. I gave false names. And when asked for directions to my home, I gave the address of a shop several blocks from our house.

I was frog-matched to the shop, which I claimed was run by my parents. When we got there, the shopkeeper was shocked to learn that he had miraculously gotten a brand new nine-year-old son. He could not hide his astonishment. The bus driver and the conductor were equally stunned.

But Center was a small place. After a brief moment of amazement, the shopkeeper recognized me. He immediately revealed the identity of my parents.

I was duly reported to my parents. This time around, the "gravity" of the misdemeanor had multiplied by a factor of three. Not only had I been caught performing a dangerous activity, but I had also wasted the time of the bus driver and conductor, and worst of all, lied about my lineage.

At that point, I experienced different forms of energy. The energy forms will become more apparent when we go through the underlying physics.

Conservation of Energy

In the previous chapter, we saw that to change the state of a substance, we need to apply some form of energy to it. For example, turning water into a gas requires heat energy (assuming constant air pressure). But this is not the only form of energy in the world. Even if one has never entered a physics classroom, it is easy to conjecture energy as the capacity to do something. For example, you need some energy to lift an object from the ground. If your car gets stuck in the mud, you need some energy to pull or push it out of the mud. That energy could be in the form of a team of Samaritans who lend you a hand in removing the car from the mud. Or it could be in the form of a tractor that pushes or pulls the car out of the mud.

* * *

One way of gaining a glimpse at the physicists' conception of energy is to consider its various forms. A few examples will suffice for our purposes.

There is gravitational energy that causes an object to fall from a high level to the ground. The force that pushes it downwards is gravitational energy. There is thermal energy, which causes ice to melt or water to boil.

Other forms of energy include electrical energy generated from various sources such as hydroelectric plants; nuclear energy that comes from the fusion or fission of atoms; mechanical energy from moving objects, electromagnetic energy; chemical energy from the mixture of certain substances; light energy; wave energy from the oceans; wind energy; tidal energy from tidal waves; geothermal energy; and biomass energy.

This hopefully gives you a picture of what physicists refer to as energy. A simplistic way of thinking about energy is the physical phenomenon that creates a force of some sort. In other words, a force that can move things from point A to point B.

As physicists would say, energy is the capacity to do work of one form or another.

Energy is what enables one to walk from one point to another. It is what enables planes to fly in the air. It is what allows the frying of sausages.

Energy also exists in latent form as potential energy. Potential energy comes in many different forms. Electrical energy is found in electrical fields, chemical energy in chemicals, magnetic energy in magnetic fields, nuclear energy in atoms released during atomic reactions, and elastic energy that arises when a flexible material is pulled. There is also thermal energy, which is potential kinetic energy from the motion of tiny particles of matter.

The interesting thing about energy is that the aggregate amount in a system does not change. It cannot be created, nor can it be destroyed. However, energy can be transformed from one form to another. These three attributes constitute what the physicists call the principle of conservation of energy. It is an incredibly astonishing natural phenomenon.

For example, when you lift a glass of water and place it on the kitchen table, it gains potential energy while on the table. If it drops from the table, it gains kinetic energy as it falls down to the kitchen floor. And when it hits the floor, it transforms the kinetic energy into heat and sound energy.

Note that the screaming sound that you emit when the glass breaks into pieces does not count in the energy transformation process that we are referring to here. Nor does the energy that your spouse expends while rushing to the kitchen to find out why you are screaming. These two forms of energy are not part of the aggregate energy in the particular system we are referring to in this instance.

Back to the Nakuru Incident

I live it to the reader to imagine what came of all the energy that the bus driver and conductor expended as they frog-matched me to my false parents and the energy they used to drag me to my real parents.

In fact, all that energy coalesced into a "*nyahunyũ*," a special kind of whip that my father used on special occasions.

My father, using the *nyahunyũ*, delivered the first quanta of kinetic energy on my backside. He repeated this process several times, creating tremendous thermal energy in my body. Also, the *nyahunyũ* created enormous sound energy as it landed on my behind.

The heat from the *nyahunyũ* was also transformed into instant chemical energy. The chemical energy passed through my nervous system at lightning speed, straight into my brain, creating enormous pain. There is no such thing as pain energy, so this was a novel phenomenon beyond the realm of physics.

The pain immediately triggered other chemical reactions in my body that produced uncontrolled movements in my mouth, manifesting kinetic energy. I cried loudly, pleading for mercy and committing to never ever engage in dangerous play.

* * *

The design of upcountry buses was changed after a few years. The ladder was moved from the back to the side of the bus, denying many children of Nakuru the enjoyment of a kinetic energy experience. Nowadays, most upcountry buses carry luggage in their underbelly.

* * *

So, what has this got to do with the conservation of energy? If we consider the incident as something in a closed system, we can use physics's energy conservation laws to better understand what was happening.

The sum total of the energy from the joy ride on the upcountry bus, and the energy used as we walked to the shop run by my "fake" parents, was equal to the potential energy on my father's *nyahunyũ*, plus the heat, sound, and chemical energy that crystallized on my backside; and that subsequently manifested itself in the form of high wailing sound from my mouth. So, in a nutshell, all the different kinds of energy were conserved in my young, perpetually enjoyment–seeking body.

CHAPTER 5

Ecology

"The more clearly we can focus our attention on the wonders and realities of the universe about us, the less taste we shall have for destruction."
— Rachel Carson

I WILL NEVER FORGET THE LAST two sessions of biology that I attended in high school. The subject matter was significantly different from what we had covered in the biology class during the preceding eight semesters. And yet, the topic was so exciting and of practical importance. Even now, I can vividly see the teacher's image as he walks from the front to the back of the biology laboratory talking about ecology.

Ecology Anecdote

A famous story of a health campaign initiated by the Chinese government in 1958 vividly illustrates the importance of the balance of nature, a key aspect in the study of ecology.

Under the leadership of Mao Zedong, the Chinese government was eager to improve the health of Chinese citizens. This necessitated removing all insect pests and animals that caused diseases such as tuberculosis, plague, cholera, polio, malaria, smallpox, and hookworm.[11]

The massive multi-pronged health campaign involved vaccinations against smallpox and the plague and significant sanitation infrastructure improvements as part of the Four Pests Campaign initiative.

The Four Pests Campaign aimed to eradicate malaria-causing mosquitoes; rats responsible for spreading the plague; flies that were a nuisance; and sparrows that ravaged rice and grain fields. "These four pests – flies, mosquitoes, rats, and sparrows – were charged with public health treason and widespread irritation."[12]

The government instructed citizens to use their best endeavors to deal with the four nuisances. What followed was a display of patriotism of significant proportions.

None of the four perpetrators of disease was spared across the length and breadth of China. A well-orchestrated propaganda campaign ensured the utmost efficiency in going after the four public enemies.

Contests amongst schools and government agencies were used as one of the tactics. The winners were those who handed in the largest number of rat tails, dead sparrows, dead mosquitoes, and flies.[13]

The results were impressive. A humongous 100 million kilograms of flies were exterminated; 11 million kilograms of mosquitos were decimated; 1.5 billion rats met an untimely death, and one billion sparrows were smashed under the special Smash Sparrows Campaign.

The sparrows would be shot down, their nests destroyed, and their eggs broken. In some instances, people would loudly bang pots and pans. The sparrows would fall dead on the ground out of fatigue after being unable to rest on tree branches. [14]

After successfully eliminating sparrows or "the public animals of capitalism," the Chinese government shifted its focus to bed bugs.[15]

But the Chinese government had unknowingly tampered with the ecological balance. The consequences were stupendous. The destruction of sparrows had removed the natural predator for locusts. Accordingly, locusts bred in large numbers and devasted expansive fields of grain. Consequently, more than 20 million people starved to death between 1958 and 1962.

Upon realizing the devastating effect of the campaign, the Chinese imported 250,000 sparrows from Russia to restore the ecological imbalance[16]

The Cobra Effect

The story of the health campaign in China is an example of the cobra effect, a phenomenon denoting an action's unintended consequences.

The cobra effect is derived from a famous story of colonialists in India who wanted to eliminate large populations of cobras that had become a menace in New Delhi. Anyone who killed a cobra and produced the cobra's tail as proof was paid a bounty.[17]

The campaign appeared to be very successful, given the many cobra tails submitted for the bounty payment. But, to the astonishment of the colonialists, the number of cobras in the city increased despite increased bounty payments.[18]

Apparently, a cottage industry on the rearing of cobras had emerged as people tried hard to earn as much bounty as possible. So, rather than the campaign reducing the number of cobras, it had the opposite effect.

A More Personal Cobra Effect

Belgium produces some of the best chocolates in the world. The country is actually one of the four top producers of chocolate in the world. The other three are the USA, Germany, and Switzerland. The annual revenues from sales of Belgian chocolate are estimated at US$ 12 billion.[19]

There are about 2,000 chocolate shops in the country. Most of the chocolate is still manufactured using traditional methods. The manufacturing practices are governed by strict laws to ensure the quality of chocolate is not compromised by the use of artificial, vegetable, or palm oil-based fats. Because of this, Belgian chocolates are a must-have for anyone who visits Belgium.[20]

There are many different varieties of Belgian chocolates. Some pralines are filled with nuts, fruit creams, and the like. There are truffles, which are powdery with a solid cocoa coating. There is the Gianduja variety, which comes in small rectangular blocks with beautiful gold-colored wrapping.

However, variety is a good thing but can sometimes be deceptive. I discovered this in a rather unusual way.

Sometime around 2000, on return from Belgium, where I was based, I visited my parents at Kinoo. In Kenya, it was customary that whenever one returned to the country from overseas, they would carry some gifts for family, relatives, and close friends. And what better gift to bring from Belgium than chocolate.

As soon as I arrived at my parents' home at Kinoo, I opened a few packets of chocolate and dished the chocolate around. There were several people in the house, primarily children.

Everyone much enjoyed the chocolates, particularly the children. But, after a few minutes, my mother started complaining that some of the pieces contained a bitter liquid at the core. After a brief examination of the chocolate wrapper, I realized what was happening. I had given out several liquor-filled chocolates (chocolate liqueurs). I scrambled to retrieve that particular variety of chocolate from the children. The situation was a little bizarre and momentarily dampened the merriment of some children.

It is not very pleasant for an adult to snatch chocolate from a child, but it had to be done. Fortunately, none of the children exhibited the kind of phase transitions witnessed in my father's homestead several years earlier, as described in the Alphonse story in Chapter 4.

Back to Ecology

The ecology ideas that the teacher taught us during those two biology lessons in high school significantly impacted my worldview. It suddenly occurred to me that invisible forces controlled the different living and non-living things. Further, human beings' actions could destabilize the existing state of equilibrium.

One of the things that poignantly affected me was the realization of the inter-dependencies in nature and the factors that ensure a proper balance of life. Before that, the words "balance of nature" existed just as a cliché used by environmentalists. But I was mistaken, as the biology teacher made it abundantly clear.

I wished that many more people could learn and internalize the ideas I learned during the two biology lessons.

But what were those ideas? Perhaps the easiest way of explaining them is through an example.

* * *

As the sun rises every day, it transmits heat energy to the earth. The energy heats the oceans and other water bodies. The process of evaporation from the water bodies creates moisture in the atmosphere that turns into clouds. The clouds are then blown by the wind in different directions.

Some clouds condense into moisture, which falls on earth as rain. The rain falls on plants and also on the soil.

The soil absorbs the moisture and other inert elements, such as nitrogen, to form nutrients. The plants absorb the nutrients from the ground using fibers (roots) and convert them into foliage.

The foliage is then eaten by animals. The animals are eaten by other animals. If the animals do not get enough nutrients, they perish and provide nutrients to microorganisms that exist in the soil.

Different species of plants and animals are engaged in a never-ending competition for food. Those species that are unable to compete effectively may starve to extinction. This phenomenon is indeed at the heart of the evolution process of "survival of the fittest" that Darwin (2008) described so eloquently.[21]

The process of ecology described above happens in detail at different levels in different ecosystems, as illustrated in the following example provided by Frenay (2006):[22]

Mushrooms compete with puffballs, seedlings, flowers, and slime for a footing in the moist rot of the old tree's body. Beetles tunnel toward its heartwood, opening a path for mites and nematodes. Termites gnaw on it as parasites in their guts digest it for them. Slugs inch their way across the carpet of the moss while crickets leap from place to place overhead. A salamander darts for a worm as, in great lurching gulps, a frog devours a butterfly. Woodpeckers land and start digging for ants. A snake glides under loose scraps of bark, not far from where a mouse and her young huddle in a shady gap.

The human population is part of the ecosystem and is subject to its own set of interesting dynamics. However, unlike the other species on earth, human beings have developed the capacity to interfere with the ecosystem's natural rhythm in both positive and destructive ways.

Human beings have developed the technology to grow food on a large scale and medicines to treat diseases – bacteria and viruses that would otherwise harm human beings. Accordingly, the human population multiplied from 3.0 billion people in 1960 to 7.7 billion in 2019 (more than double). See Figure 6.[23]

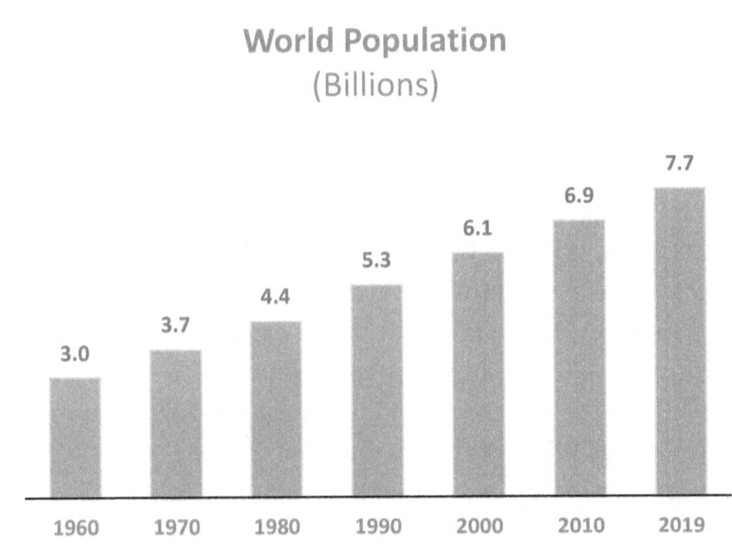

Figure 6: World population based on World Bank statistics

The increased population has been accompanied by rapid industrialization. This has, in turn, adversely affected the environment due to air and water pollution, destruction of forests, increased urbanization, global warming, and pressure on existing resources. For example, although 75% of the earth is covered with water, only 3% of the water is fresh and can be used for consumption. Not only that, approximately 99% of this freshwater is underground or in the form of glaciers and ice caps.[24] We may not experience the adverse effects of these factors directly as individuals. Still, the impact on the entire human community is becoming increasingly evident and worsening by the day.

By studying populations of different species, ecologists have noted certain interesting trends in various living communities of organisms. For example, depletion of food for a particular species of living things may result in a switch in the organism's consumption patterns – for instance, feeding on new plants or animals. This switch may cause a cascading effect in the food chain and could result in species extinction if something is not done to check the trend.

Another example is the over-exploitation of the sea by humans. This could cause a depletion of certain species of fish. The predators of that species of fish could, in turn, face starvation and, eventually, become extinct.

Species could also exhibit a rebound effect. This occurs when members of a species die out because of challenges of obtaining food, leaving behind more robust species that can withstand the challenges. An excellent example is the *staphylococcus aureus* bacteria, which was affected by *penicillin*. Over time, *it* developed strains resistant to *penicillin*, necessitating the development of more potent varieties of *penicillin*.[25]

The critical point to note is that nature seems to operate according to certain rules. The non-observance of these rules can have unexpected consequences. Sometimes the results can be uncontrollable chaos.

PART 2

Basic Ideas in Mathematics

CHAPTER 6

The Theory of Limits

"All limits are self-imposed."
—Icarus

A SIGNIFICANT PORTION OF CALCULUS, which some people find abhorrent, is founded on the theory of limits, a straightforward concept, but a potent one. Some of the offshoots of the theory of limits are mind-bending.

The Theory of Limits

There is one fundamental idea around the theory of limits: if you cannot measure something to its absolute accuracy, you can get an excellent approximation by examining what happens at the extremities.

Suppose you wanted to measure the length of a curve (or arc) such as the one shown in Figure 7. One of the best ways of estimating it is to divide the curve into multiple segments using lines, as shown in Figure 8. And by measuring the lengths of the lines (the dotted lines) and totaling all the lengths, you can get a good approximation of the length of the arc.

Figure 7: Curve (or Arc)

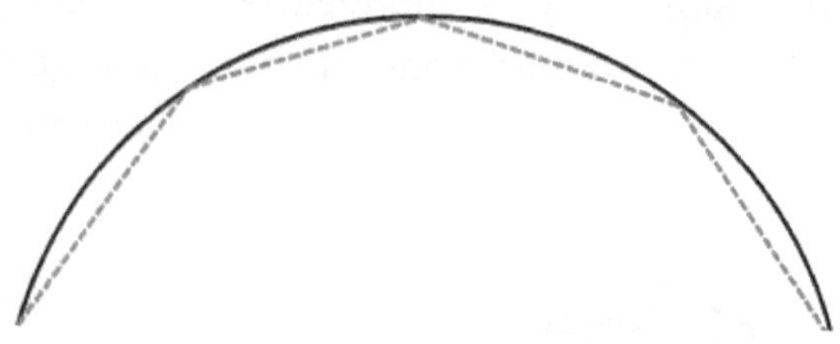

Figure 8: Curve divided into 4 segments

The higher the number of segments you can create and measure, the closer you will get to the curve's exact length.

To put it in mathematical language, the length of the curve is equal to the aggregate total of the lengths of the dotted lines, as the number of dotted lines tends to infinity.

Another way of looking at it is that the length of the arc is equal to the aggregate total of the dotted lines' lengths as the length of each dotted line becomes infinitely tiny (or as the length of each dotted line tends to zero).

If you are a non-mathematician and have followed the line of reasoning, you can stop here briefly. Take a deep breath, take a sip of your favorite beverage, and congratulate yourself for gaining an excellent understanding of one of the basic tenets of differential calculus. It is no mean achievement.

On the other hand, if you feel much enthused, you can call one of your close friends to tell them that you have just become an expert in infinitesimal calculus.

Obviously, there is more to it, but we do not want to go beyond the essence of it, as described above

* * *

The principles of infinitesimal calculus are used extensively in mathematics to find answers to exceptionally complex problems. Indeed, the degree to which mathematicians have taken these ideas is mind-boggling. That is why you may hear some people say that mathematicians live in another world of their own.

Warning

Please Google for the calculation of the length of an arc using calculus at your own risk. What you will see is not for the faint-hearted. In fact, if you are the bungy-jumping type and are ready for anything, then be my guest, but do not say you were not warned. I only hope you will not end up in one of the Websites of mathematical obfuscators who take immense joy in utterly confusing lesser mortals. However, if you have the courage, you can have a sneak preview in the House of Maths Horrors in the Appendix.

CHAPTER 7

Benford's Law

"Mathematics is the language with which God has written the universe."
—Galileo Galilei

I WAS ASTOUNDED WHEN I FIRST encountered Benford's law. The law shook the foundations of the comfort I had gained over the years about the practical applicability of the theory of probability.

Suppose someone told you that several discrete entities have an equal chance of appearing in the world without external intervention. If you were an ordinary mortal, you would most probably know intuitively that the probability of seeing any of the items would be exactly the same.

Say you are given four similar-sized balls of different colors (white, red, black, green). You are then asked to put them in a closed container with a lockable outlet through which only one ball can pass. After putting the balls in the container, you are asked to shake it. Further, you are asked to let out the balls, one at a time.

It should be evident that the chance of the first ball to exit the container being white is exactly the same as for any other ball color. In other words, there would be a 25% chance of getting a white ball, a 25% chance of getting a red ball, a 25% chance of getting a black ball, and a 25% chance of getting a green ball. This is a straightforward conclusion. The same reasoning will apply if you put a smaller or a larger number of balls into the container. You do not need a degree in mathematics to see the logic behind this reasoning.

But hold your horses right there. Life is not as simple as it looks.

Benford's Law

Benford's law introduces some weirdness in probability, but mathematics gurus probably have a good explanation.

To understand how Benford's law works, let us start by posing a simple question. If you peruse any daily financial newspaper and extract all the numbers you can see (e.g., stock prices, projected temperatures for the next few days, number of people involved in a major mishap, etc.), what are the chances that the first digit of each of the numbers would be a 1? How about 2? How about 3? And so on, until the number 9?

Common sense would suggest that each of the nine digits would have an equal chance of being the first digit. In other words, each digit would have a 1 in nine chance of being the first digit (or an 11% probability for each digit).

Well, not so, according to Benford's law. The probabilities are actually as shown in Table 1.[26]

Table 1: Probabilities according to Benford's law

Digit	Probability (%)
1	30.1
2	17.6
3	12.5
4	9.7
5	7.9
6	6.7
7	5.8
8	5.1
9	4.6

This insight is used by auditors to check for any doctoring of accounting records. Tax authorities also use it to sniff for any hanky punky in tax returns.

The law was used to detect election fraud in Iran in 2009. Banks use the law to detect loan disbursement anomalies.

The law applies to large data sets that grow exponentially. It does not apply to small data sets or data sets with a natural upper limit such as peoples' heights, weights, or IQ. However, large data sets generated randomly from these sources exhibit Benford's law.[27]

A Brief History of Benford's Law

The ideas described above were first enunciated in 1881 by Simon Newcomb, a Canadian astronomer–mathematician.[28] The idea sprung in his mind when he noticed that the front pages of logarithm tables in the library of his institution were more worn out than the pages towards the end of the document.[29]

Frank Benford, an American electrical engineer and physicist, further expounded on Newcomb's ideas. In his paper, *The Law of Anomalous Numbers*, published in the Proceedings of the American Philosophical Society in 1938, Benford provided multiple examples of data sets that exhibited Benford's law. The data sets included areas of rivers, population statistics of different countries, death rates, extracts of data from newspapers, atomic weights, and addresses.[30]

An interesting aspect of the law is that it applies irrespective of the unit of measure. For example, it applies to areas of countries whether the areas are expressed as square miles, acres, square feet, or acres. It is truly incredible.[31]

Nature does indeed have a way with numbers. It is not a wonder that Galileo said that "Mathematics is the language with which God has written the universe."

CHAPTER 8

Pi (π)

"The difference between the poet and the mathematician is that the poet tries to get his head into the heavens while the mathematician tries to get the heavens into his head."
— *G.K. Chesterton*

LET US START WITH THE two most fundamental objects of geometry, a straight line and a circle. The two objects are very closely related. For example, if you cut a circle and stretch it, you will get a line. Similarly, if you fold a straight line nicely, you will get a circle.

But let us not get ahead of ourselves. We need to proceed slowly, bearing in mind that geometry is not everyone's cup of tea. But by the end of the narrative, every reader will hopefully think of it as the best cup of tea they have ever encountered. So here we go.

What is a line? The answer ought to be obvious. Three examples of a straight line are shown in Figure 9 to ensure that we are on the same wavelength.

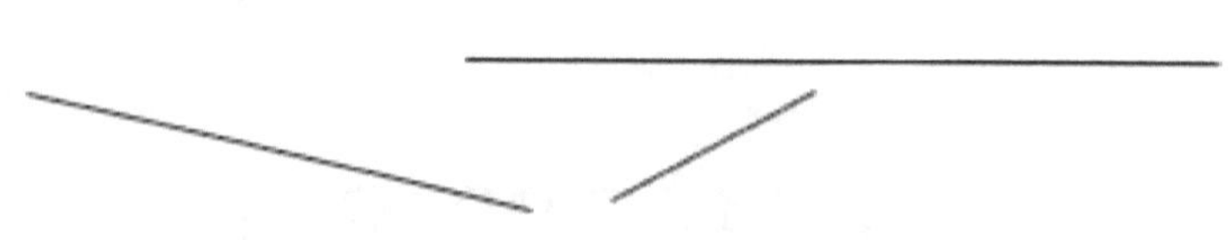

Figure 9: Three straight lines

What are some of the interesting features of lines? For a start, they do not have any curves. Secondly, they have a definite beginning and a definite end. And because of this latter feature, they can be measured with a ruler or another similar instrument to obtain their precise length.

But let us hold that final thought regarding definiteness. And here, we need to think like theoretical mathematicians, not carpenters who would not necessarily be too keen on the definiteness of a length beyond a few decimal points, say in centimeters.

Now, let us turn to the author's favorite geometrical object, a circle. What is a circle? According to the Oxford Desk Dictionary, a circle is a "round plane figure whose circumference is everywhere equidistant from its center." And once again, a picture speaks louder than a thousand words. So, we will provide three examples of a circle (Figure 10).

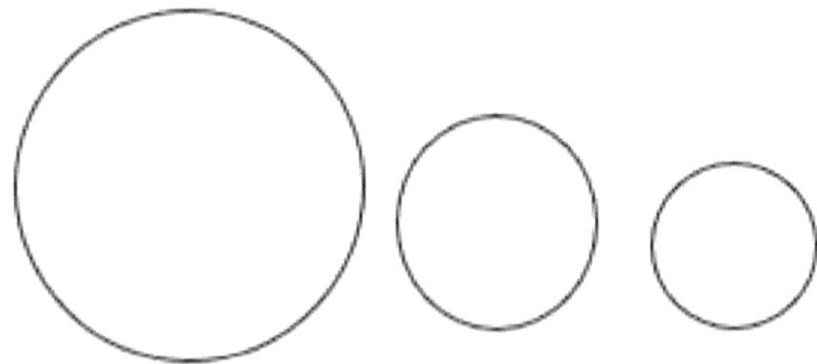

Figure 10: Three circles

Just look at the circles carefully. If you start from one point on it, go around it, and return to the same point, you will cover an exact distance in that circle. Your starting point does not matter. You will always cover the same length after taking one round around the circle.

Let us build on this idea a little further.

In theory, if you cut the circle at one point and stretch it out, you will get a straight line whose length will be exactly equal to the circle's circumference. So far, so good.

But if you think about it a little bit more and apply some of the essential high school geometry tools, things are not as straightforward as one might think.

According to high school geometry, we can calculate the circumference using a formula developed by mathematicians many centuries ago: 2 multiplied by a number called *pi* and then multiplied by the radius of the circle. This is illustrated in Figure 11.

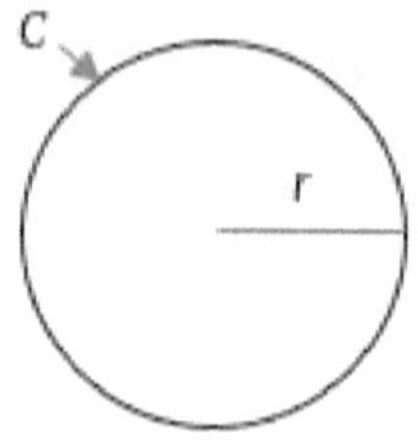

$$C = 2\pi r$$

Where:
C is the circumference
π is pi, which is <u>roughly</u> equal to 3.14
r is the radius of the circle

Figure 11: Formula for calculating the circumference

But what is the circumference to start with? If we were to cut the circle somewhere and stretch it, then the length of the string we would get is the circumference.

On the other hand, the radius is the distance from the circle's center to one of its edges.

The formula indicated above is one of the basic things one learns in high school geometry. It is the kind of formula that may be familiar to many people you may meet on the street in any urban center.

Things cannot get simpler than that. But is that so? And what is this thing called *pi*, one might ask? Here is where things become a bit muddy.

Pi is the ratio of the circumference of the circle to the diameter. In other words, pi is equal to the circumference divided by the diameter (or the radius multiplied by 2) or using the mathematical notation shown above: $\pi = \left(\frac{c}{2r}\right)$.

And mathematicians have determined that this ratio, called *pi*, is approximately 3.14.

The Dance of Lines and Circles

The ratio *pi* is 3.14, followed by an infinite number of decimal places. The number *pi* expressed to 100 decimal places is as follows:

3.1415926535897932384626433832795028841971693993 75105820974944592307816406288620899 8628034825

But why is *pi* an approximate and not an exact number? In other words, why is it that the circumference of a circle divided by its diameter is not an exact number?

On Thursday, March 14, 2019, Emma Haruka Iwao, a Japanese computer scientist and developer at Google, calculated *pi* to 31.4 trillion digits using Google cloud technology. Iwao set a new record for the largest number of digits for *pi*[32].

And the trail of numbers after the decimal point is random. There is no pattern to it.

What a remarkable thing, this *pi*? It is one of the biggest mysteries of geometry that I have encountered.

* * *

But let us pause to consider some of the implications of these simple facts about *pi*.

Firstly, it means that if you cut a circle and stretch it out, you can never really measure its length precisely. The length is 3.14 (followed by an infinite number of decimal points) multiplied by 2 and by the radius.

Secondly, what does it say about all other straight lines if you cannot get the exact length of this very straightforward line (or, more accurately, line segment)? Could it be that there is no line with a precise length?

Thirdly, if *pi* is not an exact number, it follows that even the radius of a circle cannot be exact. Here is why. If you assume that any line of any length can be converted into a circle by simply linking the two ends of the line to form a circle, then based on our discovery of *pi* described above and the formula for calculating the circumference of a circle, the radius of the circle cannot be exact. Below is an example illustrating this point, using a line 100 inches long.

Step 1: Draw a line of length 100.

Length =100 inches

Step 2: Convert the line into a circle with a circumference of 100 inches.

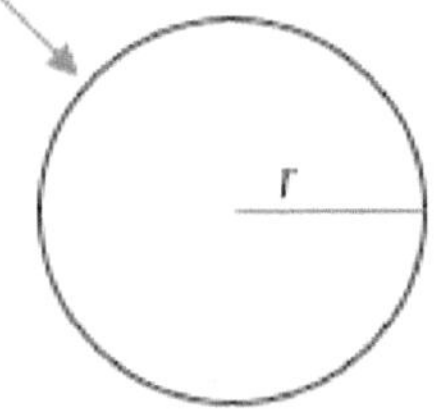

Step 3: Calculate the radius of the circle.

$$C = 2\pi r$$
$$r = \frac{C}{2\pi}$$
$$r = \frac{100}{2\pi}$$
$$r = \frac{50}{\pi}$$

This calculation shows that for an imaginary circle with a circumference (*c*) of 100 inches, the radius (*r*) is 50 inches divided by *pi* (π). And if *pi* is a number with an endless string of decimal points, then any number divided by *pi* would exhibit similar characteristics.

Another way of framing this argument is to say that if you have an exact number for *pi*, the radius cannot be an exact number and vice versa. (Something that almost sounds like one of the principles of quantum mechanics, which states that you cannot determine the location and velocity of an atomic particle. You can only determine one of these two particle attributes at any one time).

How odd can things get, and what is the underlying common-sense reason for this phenomenon?

* * *

You can visit the Maths House of Horrors in the Appendix for more interesting insights on pi, but do not say that you were not forewarned about what you might see there.

CHAPTER 9

Square Root of Two ($\sqrt{2}$)

"In real life, I assure you, there is no such thing as algebra."
— Fran Lebowitz

THE NEXT MATHS GEM THAT we will look at is the square root of two. It has almost similar roots as *pi* (pun unintended). We will use the same framework as we used for *pi* for the sake of simplicity.

The amazing thing is that it is another example of simplicity that quickly mutates into almost unbelievable complexity.

But first, a quick refresher on square roots. And to do this, we will start off with a simple table that shows a list of numbers. And for each number, we will show the square. It is straightforward and simple. And that is the first point that we will attempt to demonstrate.

Table 2: Calculating Squares of Numbers

Number	0	1	2	3	4	5
Calculation	0x0	1x1	2x2	3x3	4x4	5x5
Square	0	1	4	9	16	25

Secondly, we will work backward to calculate the square roots of the numbers.

Table 3: Calculating Square Roots of numbers

Number	0	1	4	9	16	25
Calculation	$\sqrt{0}$	$\sqrt{1}$	$\sqrt{4}$	$\sqrt{9}$	$\sqrt{16}$	$\sqrt{25}$
Square Root	0	1	2	3	4	5

In other words, the square of a number is the number multiplied by itself, while the square root is the reverse of that operation. How simple can things get?

One can think of this idea in a more pictorial way by looking at the following two examples.

A number nine can be considered a square containing nine boxes, arranged in rows and columns of three boxes each, as shown in Figure 12.

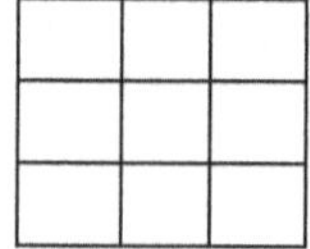

Figure 12: Three by three matrix representing the number 9

One can think of the square root of nine as the number of boxes on one row or one column of the square. In this particular example, the square root is 3.

Let us build on this idea a little further to see how it works when dealing with additions of squares.

If we start with two numbers, 3 and 4, and add their squares, then the sum of their squares is 25, as shown below.

```
3 squared =   9
4 squared = 12
Total         25
```

Using conventional mathematical notation, we can write this simple calculation as follows:

$$3^2 + 4^2 = 25 \qquad \text{or} \qquad 3^2 + 4^2 = 5^2$$

We can represent this information in the form of squares as follows:

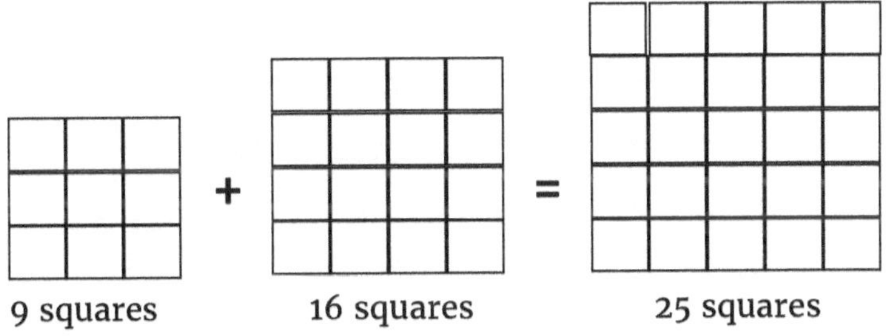

Figure 13: Illustration of the addition of squares

This is now where the beauty of simple mathematics starts to come through. You can see that [3^2 or 9 boxes + 4^2 or 16 boxes = 5^2 or 25 boxes].

The ancient Greeks approached these ideas from a different perspective, using geometry.

* * *

But first, a refresher on some basic terminology in geometry. A right-angled triangle is the type of triangle that forms a square or a rectangle when you put another similar triangle next to it, with the long sides of the triangles next to each other. See illustrations in Figures 14 and 15.

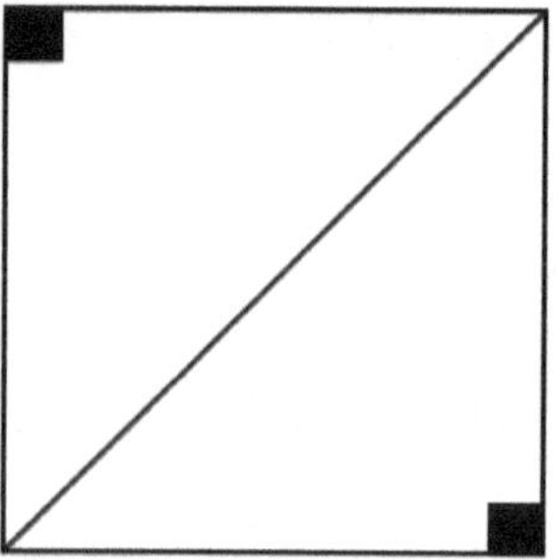

Figure 14: Two right-angled triangles next to each other

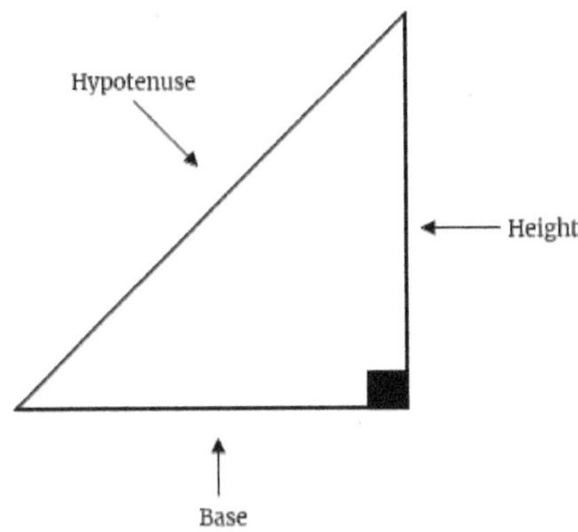

Figure 15: Right-angled triangle

The Greeks discovered a relationship between the different sides of a right-angle triangle that could be easily visualized using squares. See the illustration in Figure 16.

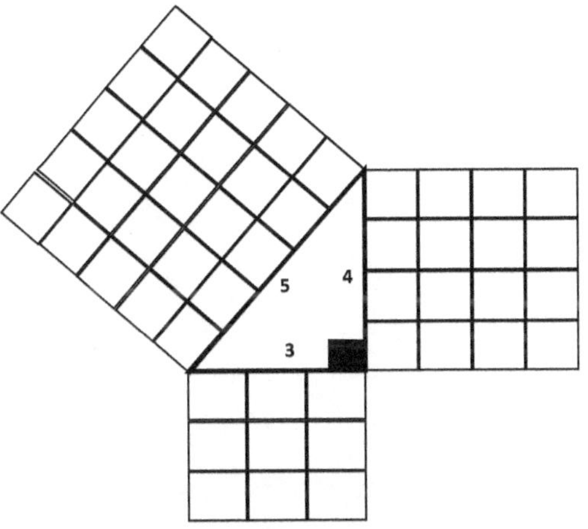

Figure 16: Right-angled triangle

Based on this picture, it is easy to see that if you have a right-angle triangle with a base of 3 and a height of 4, the hypotenuse must be 5. Or, to put it differently, the hypotenuse is the square root of 25 (i.e. $\sqrt{25}$) based on our discussion a few paragraphs earlier.

In fact, by careful study of the properties of right-angled triangles, the ancient Greeks came up with the famous formula called the Pythagoras theorem around 500- 600 BCE. This theorem states that the length of the hypotenuse of a right-angled triangle is equal to the square root of (the triangle's base squared, plus the height of the triangle squared). And in the spirit of simplicity, this magnificent equation is shown graphically in Figure 17, using the numbers from our example.

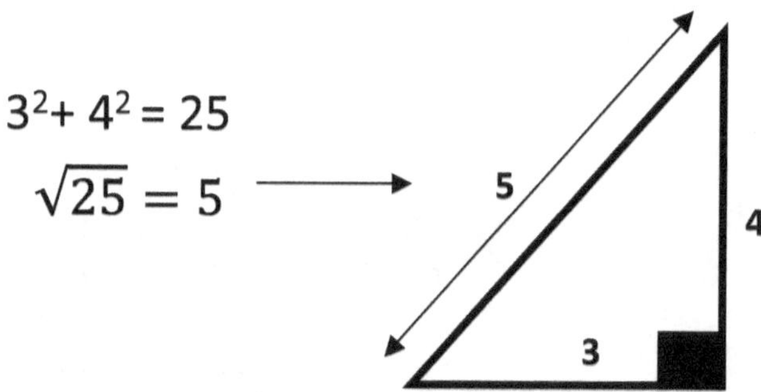

Figure 17: Illustration of Pythagoras Theorem

Legend has it that upon discovering this theorem, Pythagoras sacrificed an ox to thank the Gods for enabling him to discover the theorem.[33]

A Small Headache for the Greeks

However, there was only one nasty problem that the ancient Greeks ran into. The problem was almost similar to that of *pi* that we encountered earlier. And just for the sake of simplicity, we will use some of the geometrical objects that we used when discussing *pi* to describe the problem. To this end, let us revisit our favorite object, the circle.

If you were to draw a circle with a radius of 1, then applying the Pythagoras theorem principles, the length of the hypotenuse of the triangle shown in the diagram below would be $\sqrt{2}$. This is because $1^2 + 1^2 = 2$; therefore, the hypotenuse is equal to $\sqrt{2}$.

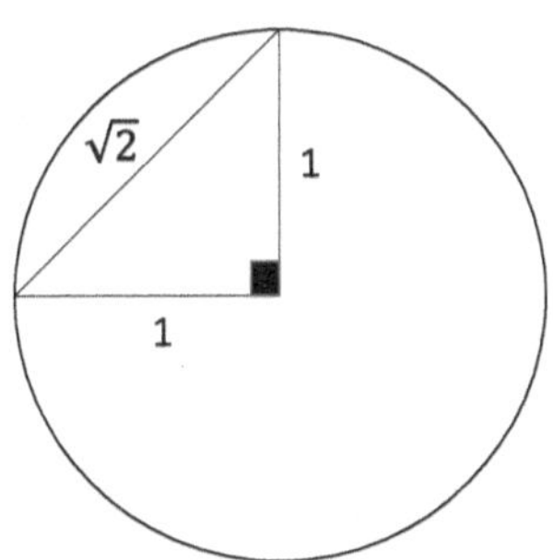

Figure 18: Illustration of the square root of 2

The triangle in the circle shown in Figure 18 is called an isosceles triangle because two sides are of equal length. This particular instance is a right-angled isosceles triangle because the angle between the two equal sides is ninety degrees.

Figure 19 is a beautiful and straightforward illustration of how $\sqrt{2}$ occurs. However, behind this simplicity is enormous complexity.

Firstly, why does it not seem to obey the Pythagoras theorem of squares? See Figure 19.

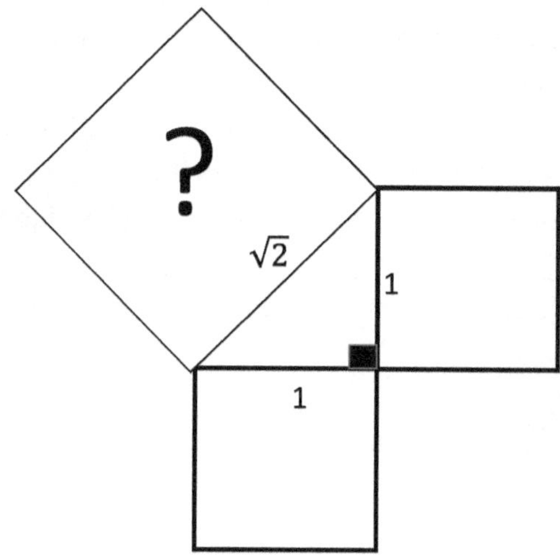

Figure 19: Right-angled triangle with a hypotenuse of $\sqrt{2}$

In other words, how can one draw boxes with sides measuring $\sqrt{2}$? This does not seem to make any sense.

Secondly, and even more interestingly, what is the value of the number $\sqrt{2}$? Once again, we enter into the realm of complexity similar to what we encountered with *pi*; namely, that $\sqrt{2}$ is the number 1.4142 with an infinite number of decimal points. Why is this?

Considering that the two sides of the triangle are exactly equal to 1, why is it not possible to get an exact number for the length of the triangle's hypotenuse?

There seems to be something very weird about certain qualities of circles and the triangles drawn in them.

* * *

On February 17, 1600, the Catholic Church executed Giordano Bruno, Italian philosopher, mathematician, and cosmologist, because he was advancing a theory about the geocentric earth contrary to the church's teachings.

It is interesting to note that even mathematicians are not immune from such heinous acts, as we will see in a minute from the story of the Pythagoreans.

* * *

The Pythagoreans were a group of what one might call fanatic mathematicians. They had a strong belief in numbers. The Pythagoreans' inner circle, referred to as "*mathematikoi*," believed that everything in the world was driven by numbers. Indeed, there was an inscription at their school's entrance that read "All is Number." All the cardinal numbers (one to ten) had a special significance to them. The even numbers (2,4,6,8,10) were considered female, while the odd numbers (1,3,5,7,9) were considered male.

They revered the Pythagoras theorem until they encountered a small nasty problem. The theorem did not hold in certain situations. For example, if one had a square with sides measuring 1, it was not possible to determine the exact length of the line crossing from one corner of the square to the opposite corner of the square (d) using the theorem (Figure 20).

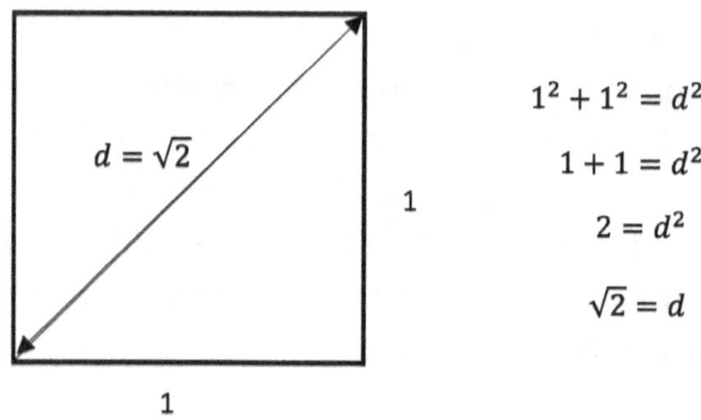

Figure 20: Pythagoras Theorem

It was Hippasus of Metapontum (c – 520 BC) who provided the proof that the square root of 2 was an irrational number, namely, a number that cannot be expressed as a ratio of two whole numbers. And by doing so, he undermined a vital belief of the Pythagoreans and indirectly signed his own death warrant. The Pythagoreans initially concealed this information because it was contrary to all the known mathematical tenets.

It is believed that the Pythagoreans committed a heinous act (almost similar to what the Catholic Church did to Giordano Bruno centuries later) simply because Hippasus revealed a dirty mathematical secret: that the square root of two was an irrational number, something that was anathema to the Pythagoreans. Hippasus was murdered by being thrown off a boat on the coast of Greece on a stormy day. [34]

* * *

The first thing that comes to many people's minds when they think of mathematics is "exactitude." Yet, here we are with evidently simple properties of circles and triangles that do not seem to make any sense whatsoever.

Could there be an approach to mathematics that has still not been discovered?

If you were to research the different ways that mathematicians have tried to explain the simple notion of $\sqrt{2}$, you would most likely be shocked (or probably go berserk if you do not have a Ph.D. in mathematics). Some explanations are included in the Appendix for those in the mood for some mathematical "gobbledygook." And please pardon me, you great mathematicians who see the hidden beauty in this punishingly complex mathematics).

CHAPTER 10

Square Root of Minus One ($\sqrt{-1}$)

"I know that two and two make four - and should be glad to prove it too if I could - though I must say if by any sort of process, I could convert 2 and 2 into five it would give me much greater pleasure."
—Lord George Gordon Byron

I F YOU ARE NOT A mathematician and felt a little weird about the idea of a square root of the number two, it was because you had not encountered something even weirder, namely, the square root of -1. Thinking about this almost twisted mathematical object is enough to give one a migraine.

Mathematicians call square roots of negative numbers imaginary numbers.

How does one even represent such a number graphically? We can use elementary algebraic equations to achieve the same purpose (hopefully).

Just consider an equation such as:
$$x^2 - 4 = 0$$

You will recall from elementary algebra that to find the value of x, we just move the 4 to the other side of the equation and then find out what x works out to as follows:

$$x^2 = 4$$

In other words, a number "x" multiplied by another number "x" gives you 4. Therefore, we can easily conclude that "x" must be 2 because 2 multiplied by 2 equals 4. So, to express this information in simple algebraic terms, we can say that:

$$x^2 = 4$$
$$x = 2$$

Using square root notation, this simple calculation works out as follows:

$$x^2 - 4 = 0$$
$$x^2 = 4$$
$$x = \sqrt{4}$$
$$x = 2$$

You will agree that this is very basic and simple algebra.

Now, using the same logic but using slightly different figures, we can see that:

$$x^2 + 1 = 0$$
$$x^2 = -1$$
$$x = \sqrt{-1}$$

But now we get stuck. Which number, when multiplied with itself, equals a minus number (minus 1 in this particular instance)? It does not make sense whatsoever.

And if you feel that way, you are justified.

Around 1740, the mathematics wizard Leonhard Euler encountered the same problem. But instead of throwing in the towel, he came up with the idea of imaginary numbers, namely, numbers that would give a negative number when squared. The first of these imaginary numbers was $\sqrt{-1}$ which Euler named i. And by this simple sleight of hand, it suddenly became possible to do algebra with imaginary numbers.

So, for example, it became possible to write equations such as:

$$2x + 3i = 0$$
$$\frac{1}{2}x - 15i = 10$$

And the manipulation of these types of equations would be done in the same way as algebraic equations, as long as one kept at the back of their mind that i was equal to $\sqrt{-1}$, or perhaps more importantly that $i^2 = -1$.

And to simplify matters even further, mathematicians came up with the idea of a complex number plane. It is probably not far-fetched to suggest that the complex plane concept was driven by a desire to figure out a way of visualizing the imaginary numbers.

Consider how we typically visualize the real numbers such as -3, -1, 0, ½, 1,2,3,4,5, 10, etc. We normally think of such numbers as points on a horizontal number line Figure 21.

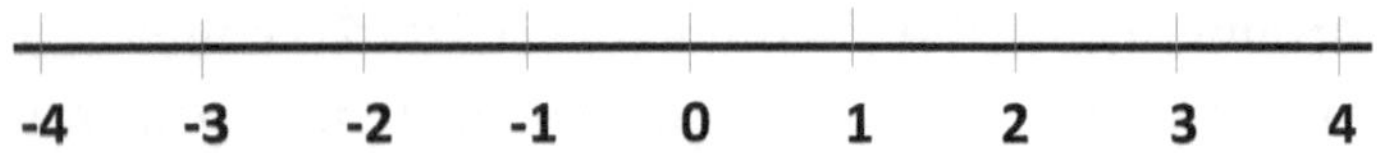

Figure 21: A segment of the number line

In the case of imaginary numbers, mathematicians came up with a vertical number line for imaginary numbers. This should not be confused with the y-axis, but it works similarly. See Figure 22.

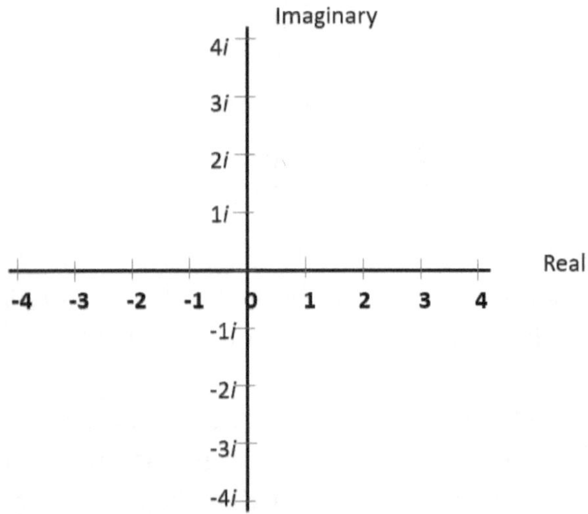

Figure 22: The complex plane

This brought about the so-called complex number theory, the subject of our next chapter.

CHAPTER 11

Complex Numbers (*i*)

"Imagination is the beginning of creation. You imagine what you desire, you will what you imagine, and at last, you create what you will."
— *George Bernard Shaw*

COMPLEX NUMBERS ARE A COMBINATION of real numbers and complex numbers. We can map real and imaginary numbers on the complex plane just as we do when working with real numbers on the "x" and "y" axis. But rather than having a "y" axis, we would have an imaginary numbers axis. See Figure 23.

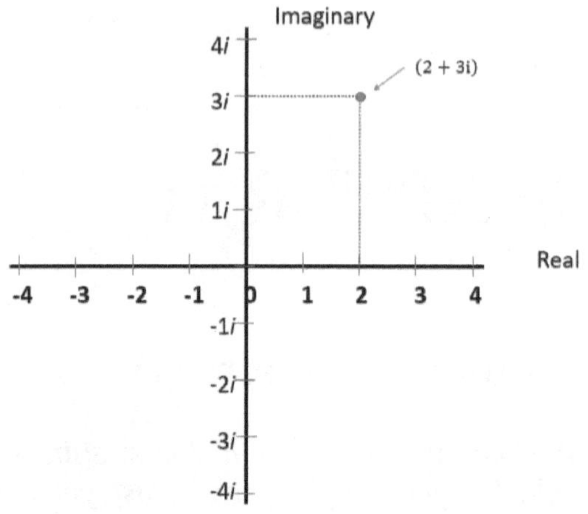

Figure 23: Mapping of one point on the complex plane

Note that whereas in ordinary geometry we would show the intersection of points on the xy plane using coordinates such as (2,3), i.e., 2 points on the x-axis and 3 points on the y-axis, in the case of complex numbers, the "coordinates" are expressed as a real number and an imaginary number. For example, (2+3i). There is no comma as in the x-y plane geometry.

Accordingly, one could classify the different types of numbers, as shown in Figure 24[35]:

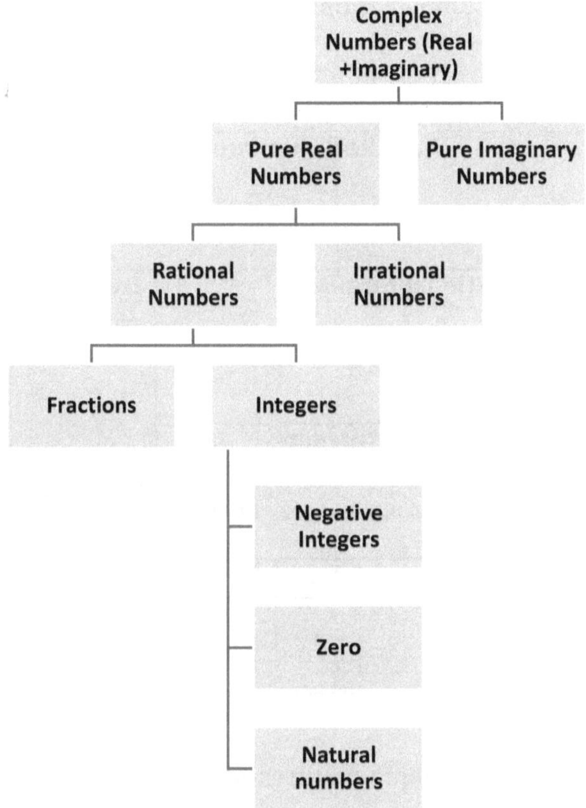

Figure 24: Hierarchy of numbers

Below is a table containing examples of the different types of numbers in the complex numbers hierarchy.

Complex Numbers (Real and Imaginary)	
1+7i , 20+8i , -34+9i, 6-2i	

Pure Real Numbers			Pure Imaginary Numbers
1, 10, -34, 6, 3π			7i, -8i, 9i

Rational Numbers		Irrational Numbers	
8, -3, 1.6, $\frac{2}{5}$		$\Pi, \sqrt{2}, e, \sqrt{3}$	

Fractions	Integers		
$\frac{1}{2}, \frac{17}{3}, \frac{8}{11}$	8, 9, 0, 3, -7, -72, -100		

Negative	Zero	Natural
-3, -7, -72,	0	1, 2, 3, 4,5

One of the unique features of imaginary numbers is that while individual numbers may not make sense, the multiplication of imaginary numbers sometimes results in real numbers. See the example below.

$$2i \times 5i = 10i^2 = 10(-1) = -10$$

This feature enables imaginary numbers to be used in many different areas of mathematics and physics. For example, imaginary numbers are used in control theory, fluid dynamics, electrical engineering, and quantum mechanics.

We will also revisit imaginary numbers in our discussion of chaos theory, an immensely fascinating subject.

CHAPTER 12

Euler's number (*e*)

"I am incapable of conceiving infinity, and yet I do not accept finity."
— *Simone de Beauvoir*

THE NUMBER *e* REPRESENTS ANOTHER mathematical mystery. It seems to make no intuitive sense to me anyway. But let us first understand what it is.

If you had $ 1 and put it in a bank to earn interest at 100% per year, then at the end of one year, your money would have increased to $ 2. In other words, the total of your original deposit of $ 1 plus interest amounting to $ 1.

If you negotiated with the same bank and they agreed to pay you interest every six months instead of at the end of the year, you would be slightly better off. You would have $2.25. The amount would be somewhat more than in the first scenario because, at the end of the first six months, you would receive an interest of $ 0.5. The interest for the next six months would include the interest on the original deposit of $1 and interest on the $0.5 that would have been credited to your account during the first six months. The same logic would apply if you received interest quarterly or monthly.

The amount would continue increasing as the duration of the periodic payments became shorter and shorter. The investment would grow to $ 2.44 if the interest was credited quarterly and $ 2.61 with monthly interest payments. See Table 4.

Table 4: Compound interest on the investment of $1

Deposit   **1.00**
Interest   **100%** per annum

	Cumulative Amount (Deposit + Interest)			
	Annual	**Semi-annual**	**Quarterly**	**Monthly**
Jan	1.00	1.00	1.00	1.08
Feb	1.00	1.00	1.00	1.17
Mar	1.00	1.00	1.25	1.27
Apr	1.00	1.00	1.25	1.38
May	1.00	1.00	1.25	1.49
Jun	1.00	1.50	1.56	1.62
Jul	1.00	1.50	1.56	1.75
Aug	1.00	1.50	1.56	1.90
Sep	1.00	1.50	1.95	2.06
Oct	1.00	1.50	1.95	2.23
Nov	1.00	1.50	1.95	2.41
Dec	**2.00**	**2.25**	**2.44**	**2.61**

However, quite amazingly, the increase would continue to become smaller and smaller until the incremental amount would become minuscule.

For example, if the periodic interest payments were made daily, the investment would increase to $ 2.71457. It would grow to $ 2.71813 if interest was credited hourly.

In fact, in 1683, the mathematician Jacob Bernoulli discovered that irrespective of the number of installment payments, the amount would always converge to a number starting with 2.7

This number is what mathematicians call Euler's number, or simply *e*, in remembrance of the renowned Swiss mathematician Leonhard Euler who began using it in his publications regarding the explosive forces of cannons around 1727/28.

Similar to π, the number *e* has an infinite number of decimal points. The number written to forty decimal places is as follows:

2.7182818284590452353602874713526624977572

Why does the value of *e* behave in this odd manner? Once again, nature seems to throw a curveball at us. But this only makes us even more inquisitive.

An Even Greater Peculiarity

There is an even more remarkable peculiarity about *e* that is perhaps one of the great marvels of mathematics. I talk about it so often that I sometimes feel guilty of monotony.

It is the discovery by the mathematical wizard Leonhard Euler of the unique relationship between e, i and π, what mathematicians call the Euler Identity. It is expressed as follows:

$$e^{i\pi} - 1 = 0$$

Here is what Keith Devlin, a mathematician from Stanford University, said about this equation in 2002:

Like a Shakespearean sonnet that captures the very essence of love or a painting that brings out the beauty of the human form that is far more than just skin deep, Euler's Equation reaches down into the very depths of existence.[36]

Euler's identity has been described as the most beautiful formula in mathematics, notably for its conciseness and the fact that it captures the relationship between five fundamental constants in mathematics. Firstly, e, which we saw earlier, occurs naturally in the analysis of compound interest. Secondly, the number i, or the square root of minus 1 (i.e. $\sqrt{-1}$), the most fundamental imaginary number we encountered earlier. Thirdly, π, the ratio of the circumference of a circle to its diameter. Fourthly, the number one (1), the first and smallest positive integer, and the first in the sequence of natural numbers. And finally, the number zero (0), the central number in the number line.

I believe everyone needs to know about Euler's identity, but its derivation clearly belongs to the Maths House of Horrors.

A Snapshot of Euler's Biography

Leonhard Euler was an exceptional human being by many accounts.

He was born in Basel, Switzerland, in 1707. He joined the University of Basel at the age of 13. He obtained a Master of Philosophy degree while still a teenager. He was initially destined to become a clergyman, but his private tutor Johann Bernoulli (1667 – 1748), a prominent mathematician, persuaded Euler's father that Euler should pursue a career in mathematics.[37]

In 1727, Euler moved to Russia. He joined the St. Petersburg Academy. Initially, he worked in the medical department. He was subsequently promoted to the mathematics department. He quickly moved up the ranks and became a professor in 1731, at the age of 24.[38]

He married Katharina Gsell in 1734. They had 13 children, but only five of them survived early childhood.[39]

He moved to Berlin in 1741 and joined the Berlin Academy. It was while in Berlin that his intellectual prowess really came through. He published hundreds of articles in physics and mathematics, including ground-breaking works in calculus. His name appears almost everywhere in physics and mathematics. His contributions in geometry, trigonometry, calculus, differential equations, number theory, and notational systems were revolutionary, not to mention breakthroughs in astronomy/lunar motion, acoustics, mechanics, and music. And these are just a few of the numerous fields in which he made ground-breaking contributions. The enormity of his contributions is almost beyond belief.[40]

How is it that that one person could have such a huge intellect? This is a phenomenon that needs to be studied in detail. Perhaps there is an insight here that awaits discovery.

CHAPTER 13

Infinity (∞)

"If the doors of perception were cleansed, everything would appear to man as it is - infinite."
— William Blake

WHAT IS INFINITY? THE ANSWER to this question should be obvious. It simply means boundless. In other words, something that has no end. For example, if you count the natural numbers 1,2,3,4,5, etc., you can continue counting forever without reaching the final number. And this is a straightforward idea to grasp.

However, things become a bit murky if you examine the concept of infinity a bit more closely. For a start, is infinity a single point somewhere in the distant horizon? The answer is clearly "No" because if you start counting the natural numbers, as described above, you can never really reach a point where you exhaust the number of digits. There will always be an additional number to count. The mathematical symbol denoting infinity is "∞."

Having said that, one can state with certainty that there are different types of infinities. For example, some infinities are larger than others. This is easy to demonstrate.

Consider the counting numbers. You can obtain a larger number for each digit by multiplying it by another positive number. So, suppose you multiply all-natural numbers by two. In that case, the new number series will be larger than the original one by a factor of two, including the numbers in the infinite horizon.

Infinity can also be perceived in terms of smallness rather than largeness. It is called infinitesimal. Take the number 1. You can divide it into smaller numbers by dividing it by the counting numbers, namely, $1 \div 2$, $1 \div 3$, $1 \div 4$, $1 \div 5$, $1 \div 6$, etc. Each successive number is smaller than the preceding one. And clearly, you can never get a number that you could say is the smallest. The smallness runs into infinity.

But despite the impossibility of determining specific numbers in infinities, it does not mean that we cannot do mathematical calculations using the concept of infinity. Indeed, the branch of mathematics called calculus is hinged on the idea of infinity. For example, one can determine the area of a circle by dividing it into infinitesimal wedges (tiny triangles from the center of the circle) and adding up the areas of each of the wedges as the number of the wedges tends to infinity.

As the numbers approach infinity, you can never reach a point where you finish counting.

But why should we be interested in infinity? The answer is that, similar to complex numbers, it lays a good foundation for our upcoming chaos theory discussion.

CHAPTER 14

Chaos

"That small difference made all the difference."
—Johnny Rich

I WAS MESMERIZED WHEN I first came across chaos theory about thirty years ago. I was convinced that it would become the most significant thing in science in the next few years. In fact, I feel disappointed that the theory did not seem to gain as much traction as I had assumed. But perhaps it was one of those things that emerged too far ahead of their time, and maybe the theory will erupt in a big way in the next few years.

What fascinated me the most was the simplicity of the theory, and its amazingly massive implications, at least on the face of it. Indeed, the theory can be encapsulated in the following words of Terry and Gaiman (2019)[41]:

It used to be thought that the events that changed the world were things like big bombs, maniac politicians, huge earthquakes, or vast population movements, but it has now been realized that this is a very old-fashioned view held by people totally out of touch with modern thought. The things that really change the world, according to Chaos theory, are the tiny things. A butterfly flaps its wings in the Amazonian jungle, and subsequently, a storm ravages half of Europe.

The original ideas around the butterfly effect, a cornerstone of chaos theory, are attributable to a mathematician and meteorologist named Edward Lorenz.[42]

Lorenz was working as a meteorologist at the Massachusetts Institute of Technology. At the time, weather forecasting was essentially a mathematical exercise. It involved feeding certain types of data to computerized prediction models and generating projections of likely weather conditions at a future date.

It does not require a lot of imagination to realize how difficult it was to make weather predictions irrespective of the mathematical models involved. It was probably easy to predict the temperature in a specific localized area in the next hour based on prevailing temperature readings, humidity, air pressure, wind speed, wind direction, and the like. But it was near impossible to predict what would happen in a week.

The truth is that meteorologists were trying hard to do something near impossible. Still, nobody was ready to openly admit it.

However, for Lorenz, computers had created an opportunity to glean into the mystery of weather patterns. Lorenz did some improvisation to see the computer output in a rudimentary graphical form. By keying in different variables in his computer model, he could graph the air currents so that he could easily recognize the cycles of air blowing from one direction to another. He noticed that the cycles displayed a certain orderly pattern, but strangely, none of the patterns repeated themselves. It was as if there was certain hidden magic behind these patterns waiting to be discovered.[43]

The breakthrough occurred by sheer chance one day in the winter of 1961. Lorenz had run his model but wanted to examine it in more detail. However, since he had already generated his model's printouts, he did not want to start everything from scratch. He decided to pick a data point mid-way through the run. Accordingly, he picked a number from the printout and keyed in 0.506. The actual number stored in the computer program's memory was in six digits, 0.506127, but Lorenz did not believe that this would be of any significance since the difference between the two numbers (0.000127) was minuscule.

After keying in the number, Lorenz walked away from the computer to get a cup of coffee. When he returned an hour later, he was astonished by what he saw.

Instead of the graph from the new run duplicating the original run's output, the pattern generated had started diverging from the original pattern, and the resemblance had disappeared entirely after several data points. And yet, the computer program generating the output had not changed one single bit.

After a further detailed review of the data and extensive contemplation, Lorenz realized that just a minor tweak in a system's initial conditions created significant changes in the outcome.

The discovery by Edward Lozenz had far-reaching consequences. Some of the predictions the meteorologists were making were extremely sensitive to initial conditions. And given the enormous challenges of accurately determining the existing weather conditions, it meant that making long-term predictions of weather conditions was almost meaningless. This was obviously quite unsettling to other meteorologists, but it was the honest truth.

This discovery led to the emergence of the metaphor cited at the beginning of this story that the tiny air currents caused by a flap of a butterfly's wings could translate into a major storm in a distant land.

This metaphor has been used to describe different types of chaotic situations. Parish (2019) gives interesting examples. The plutonium bomb dropped in Nagasaki, Japan, on August 9, 1945, by a US B29 bomber was initially intended to be dropped on the Japanese city of Kokura, where there was a munitions factory. However, due to cloudy weather conditions, Major Charles W. Sweeney could not obtain a suitable opening in the sky to drop the bomb on Kokura. After circling the city a few times, he decided to drop the bomb in the nearby city of Nagasaki. So, through the split–second decision of Major Sweeney, the residents of Kokura were spared a major catastrophe.[44]

The second example is the assassination of Archduke Franz Ferdinand on June 28, 1914, by a Serbian teenager called Gavrilo Princip. The interesting thing is that Gavrilo and two of his accomplices had made an attempt on the Archduke's life a short while earlier by planting a bomb on the road. The bomb exploded underneath the Archduke's vehicle, but he was not harmed. The driver was instructed to take a different route from the previously planned route, but he seemed not to have heard the instructions, driving straight into the assassins' line of fire. The death of the Archduke and his wife triggered the First World War. The war would probably not have happened if the driver had followed instructions. Indeed, the second world war would probably not have occurred either.[45]

* * *

The butterfly effect becomes even more intriguing as one gets a little deeper into it.

When I delved a little deeper into the subject of chaos, the essence of the simplicity and amazingly complex idea of chaotic phenomenon hit me like a bolt of lightning. At first, the concept's abstractness seemed just like another interesting idea that Edward Lorenz had come up with. However, when I experienced the reality of the phenomenon, I was utterly mesmerized.

My sense of wonder started when I read about the workings of the Lorenz wheel. It is quite a simple idea that is best explained graphically. See Figure 25.[46]

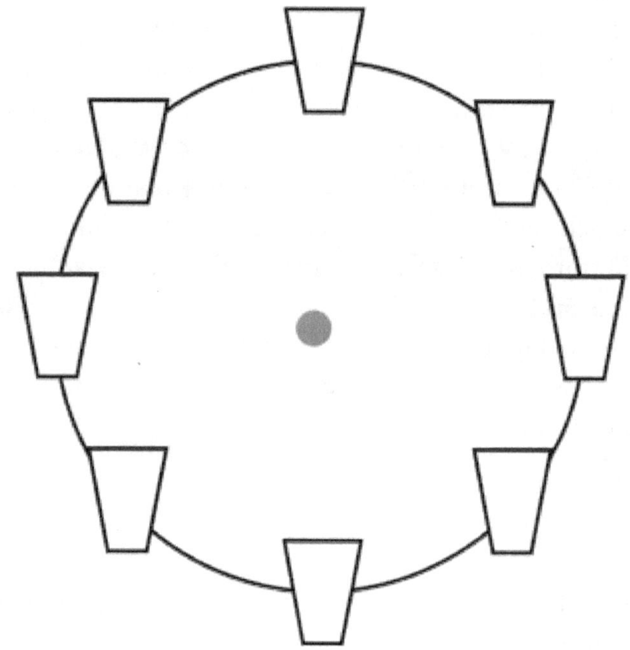

Figure 25: The Lorenz wheel

The Lorenz wheel has eight containers attached to it. All containers are exactly the same size and have holes at the bottom. There is a pivot in the middle of the wheel around which it turns with minimal friction.

Water is then poured into the containers continuously, as illustrated in Figure 26.

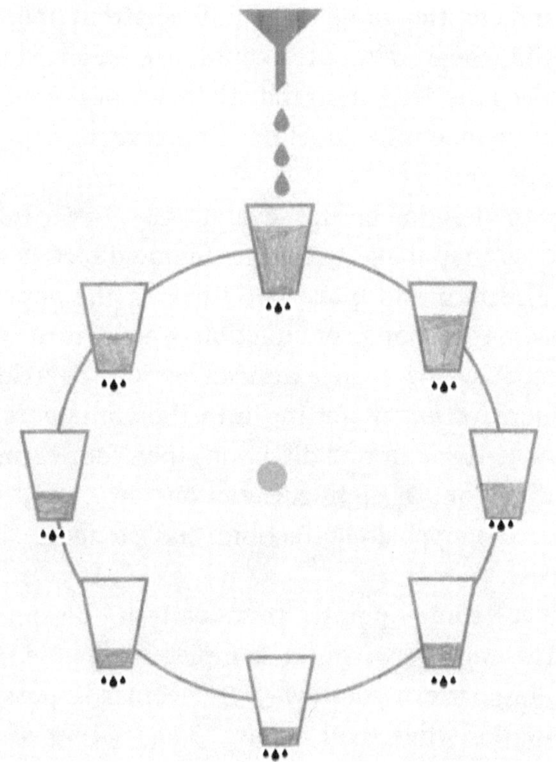

Figure 26: The Lorenz wheel in motion

As the water starts getting into the first container, some of it leaks out. Still, the weight of the remaining water forces the wheel to start turning.

And as the wheel continues turning, the next container in the wheel passes below the water source and gets a little water into it as well. This increases the wheel's motion, depending upon the rate at which water is poured into the system. As the process continues, each container gets some water into it and empties some through the bottom holes.

At some point, the weight of the containers on one side of the wheel exceeds the weight on the other side of the wheel, and the wheel changes its direction. The reverse happens after a few more turns.

The change in direction occurs several times. In the initial stages, the wheel may make two turns in one direction and then change direction and make two turns in the opposite direction. Indeed, this change of direction in equal turns may happen continuously over an extended period, depending upon the amount of water getting into the containers. In other words, four turns in one direction, then four turns in the opposite direction. Or eight turns in one direction, then eight turns in the opposite direction, and so forth, in a doubling process.

However, at some point, this pattern disappears completely. The motion becomes completely chaotic. The periodic doubling pattern goes away. It becomes impossible to predict how the wheel will behave. The motion of the wheel in the two directions becomes completely random. The pattern becomes infinitely different.[47] A most perplexing phenomenon indeed.

This phenomenon must be a nightmare for those who study fluid dynamics.

One would have thought that mathematicians and physicists would have found a formula for predicting the wheel's motion by now. But as far as I am aware, such a formula does not yet exist. It is truly incredible.

* * *

The mysterious chaotic phenomenon observed when the flow of water increases in the Lorenz wheel system amazingly occurs in other aspects of nature.[48]

Chaos applies to many different types of phenomenon, but we will stick with chaos in populations to simplify matters.

Ecologists who study the changes in populations of species are familiar with this phenomenon. However, before Lorenz's discovery, introduced in biology by James Yorke in 1972, biologists thought that the unpredictable populations, as the population growth rate increased, were due to disturbances in the data (perturbations). Some biologists discreetly avoided introducing discussion of the unusual behavior in their analysis of population trends. This idea is well described by Gleick (1988).

Non-linear systems with real chaos were rarely taught and rarely learned. When people stumbled across such things – and people did – all their training argued for dismissing them as aberrations. Only a few were able to remember that the solvable, orderly, linear systems were the aberrations. Only a few, that is, understood how non-linear nature is in its soul.[49]

A classical equation, the "logistic difference equation," was used to estimate population changes based on the reproductive rate, predators in the environment, availability of food, competition amongst the species, starvation, and so forth. According to this equation, if you started off with a population of "x" at a particular point in time, then at the next point of time, the population would have grown to r x (1-x), where "r" was the rate of population growth.

Suppose you start off with a population of 0.02 (where 1 represents the maximum population of the species in a particular environment) growing at a rate of 2.5. At the next discrete time interval, the population will be 2.5 multiplied by 0.02 multiplied by (1-0.02) or [2.5 x 0.02 x 0.98] =0.049. By repeating this calculation using the new population of 0.049 as the starting point for time 3, you will get a population of 0.116. See Figure 27.

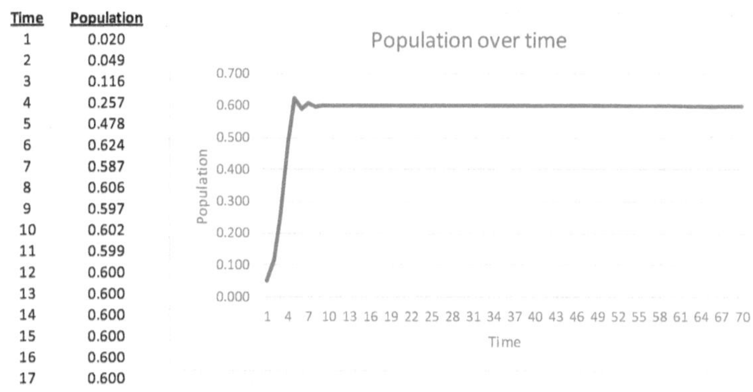

Time	Population
1	0.020
2	0.049
3	0.116
4	0.257
5	0.478
6	0.624
7	0.587
8	0.606
9	0.597
10	0.602
11	0.599
12	0.600
13	0.600
14	0.600
15	0.600
16	0.600
17	0.600

Figure 27: Population growth at a growth rate of 2.5

The first obvious observation is that the population increases quite rapidly for the first ten time periods, reaching a peak of 0.606. After that, it drops slightly and then miraculously remains at a level of 0.600.

However, if the growth rate increases to a higher number, something very interesting happens. The increase and decrease of the population will initially settle at two levels, as depicted in Figure 28.

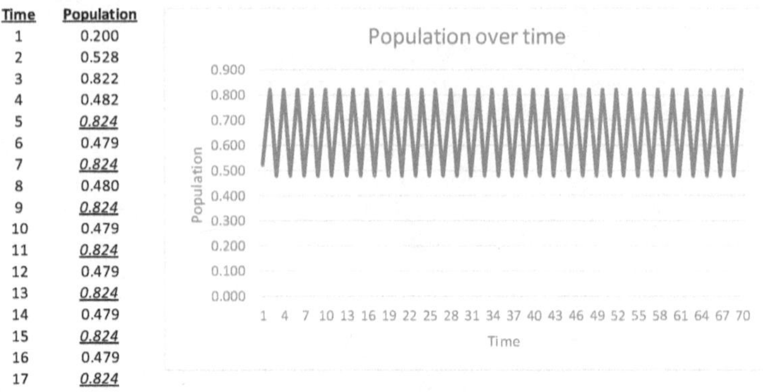

Time	Population
1	0.200
2	0.528
3	0.822
4	0.482
5	*0.824*
6	0.479
7	*0.824*
8	0.480
9	*0.824*
10	0.479
11	*0.824*
12	0.479
13	*0.824*
14	0.479
15	*0.824*
16	0.479
17	*0.824*

Figure 28: Population growth at a growth rate of 3.3

This unusual phenomenon does not end there. As the growth rate increases, the population settles at four points. And as the growth rate goes up, it settles at eight points, and so forth.

However, after a certain growth rate, something magical happens. The population level becomes erratic. It does not settle at any particular level. The pattern becomes chaotic. See Figure 29.

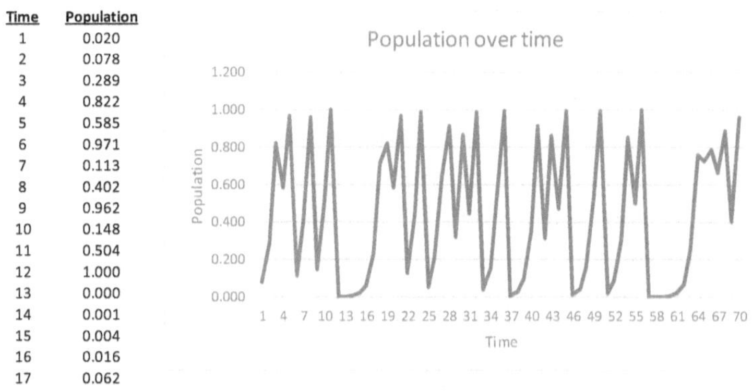

Time	Population
1	0.020
2	0.078
3	0.289
4	0.822
5	0.585
6	0.971
7	0.113
8	0.402
9	0.962
10	0.148
11	0.504
12	1.000
13	0.000
14	0.001
15	0.004
16	0.016
17	0.062

Figure 29: Population at a growth rate of 4.0

Figure 29 can be correctly described as a nightmare for ecologists.

The irony about this situation is that in many instances, the chaotic situation depicted in Figure 29 is the physical reality in the world. In contrast, the steady-state shown in Figure 27 is the aberration. By understanding the underlying drivers of the chaos in Figure 29, we are better positioned to understand the reality of the world.

* * *

Notice that the situation described above is similar to what we encountered when discussing the Lorenz wheel.

The Logistic Map

If one plots a graph showing the population growth rates on a horizontal axis and the population inflection points on the vertical axis, something even more interesting emerges. The graph, known as the Logistic Map, displays a branching out of the population levels, just like a tree. People who study chaos refer to this branching out as bifurcation, a fascinating phenomenon in nature, as we shall see a little later. Figure 30 shows a simplistic example (not to scale) of the bifurcation.

The bifurcations occur when the growth rate is equal to 3.0, 3.45, 3.54, 3.564, and 3.569.

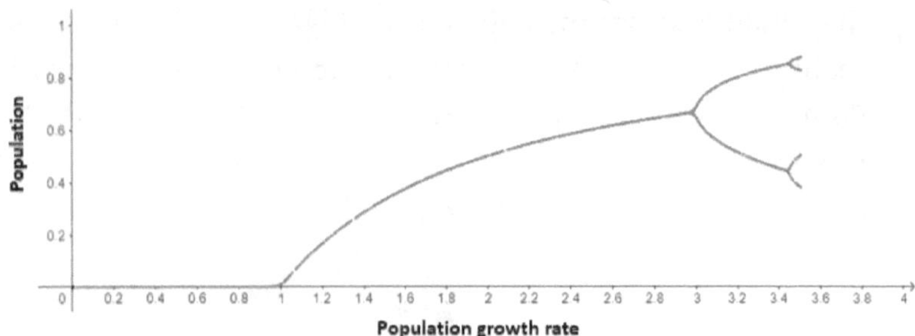

Figure 30: Bifurcation at growth rates of 3.0 and 3.45

The periodic doubling phenomenon is quite astounding. Why does it happen?

The Emergence of Chaos

But there is something even more astounding. After a certain point, the period-doubling ceases and is replaced by chaos. The forking is replaced by multiple data points that completely blur the graph after a growth rate of 3.57. The system becomes chaotic.[50] See Figure 31.

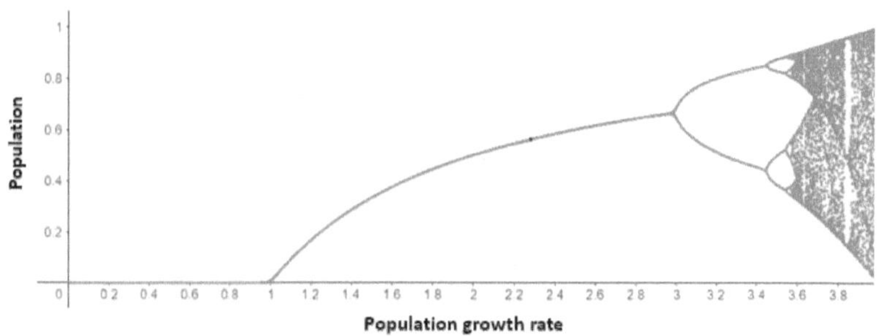

Figure 31: Emergence of chaos after a growth rate of 3.57

Chaos offers mathematicians new lenses of viewing the patterns inherent in the natural world.

The Cycle of Chaos and Stability

And then something even more amazing happens. After reaching a growth rate of around 3.64, stability re-emerges. The chaos disappears and is replaced by stability, where the period doublings occur systematically.[51][52] See Figure 32.

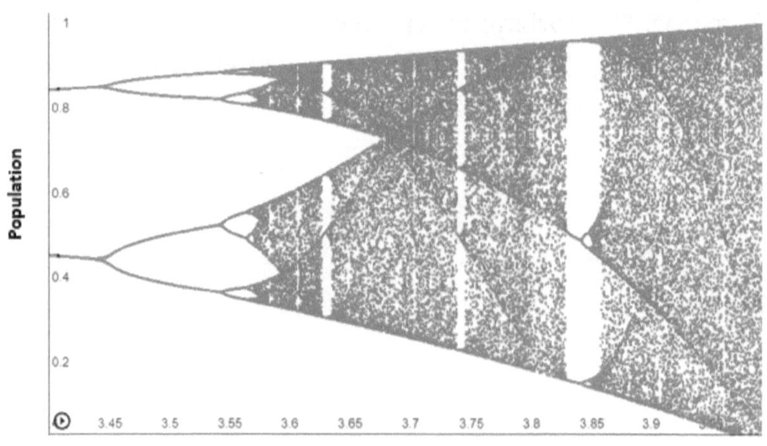

Figure 32: Stability amidst the chaos

However, stability does not last for long. The chaos re-emerges again after a growth rate of around 3.66. See Figure 32.

The shift from stability to chaos repeats itself multiple times. The cycles continue to infinity.

What is the reason for these unbelievable changes in population? I personally find the changes quite mesmerizing. It is as if there is certain magic behind the numbers.

So far, I have not found any comprehensible explanation for the phenomenon. It remains a miracle that cannot be clearly explained using existing human language.

Or perhaps more interestingly, it suggests that some vistas of mathematics have still not been discovered.

Feigenbaum Constants

The weirdness of the phenomenon described above does not stop there.

A physicist named Mitchell Feigenbaum looked closely at the period doublings described above. He noticed something extraordinary. When he measured the periodic doublings' durations and examined them closely, he noticed something interesting. Every successive period doubling was larger than the preceding period-doubling by a ratio of 4.66920 to thirty decimal points.

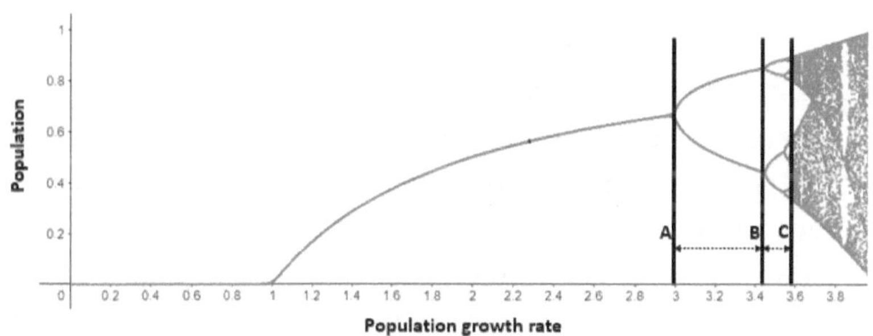

Figure 33: Relationship between bifurcation tines

In other words, if you took the period from A to B shown in Figure 33 and divided it with the period B to C, you would get 4.6692.

If you repeated this exercise at every subsequent period-doubling, you would get the same number: 4.6692. How weird can things get?

This ratio is denoted with the symbol δ by mathematicians. And just like π, the number δ has an infinite number of decimal points.

Why is this number equal to the constant 4.66920160910299067185320382157 8... and not another number?

This number 4.6692 is named after Fiegenbalm himself. It is called the Faigenbaum constant.

It is one of the many renowned mathematical constants, such as the *Euler* constant, *pi*, and others.

Some people consider the Feigenbaum constant as the symbol of universality as it represents a universal phenomenon that is manifest in different aspects of nature.

If you reflect on where we started in this discussion, you will undoubtedly marvel at the "wonder" that emerges from these few simple numbers.

Feigenbaum started examining perturbations using the equation used in elementary, high school mathematics for drawing parabolic shapes, namely, y = r x (x–x²). The same formula that was used by Lorenz.

Please do not be put off by this equation if you are not an algebra or geometry buff. All that you need to consider is the different graphical shapes produced by the equation using different starting populations (x) and projected growth rate (r), just as we did when discussing Lorenz's experiments.

For Feigenbaum, he used the formula to generate cobweb maps. These maps show more clearly the iterative manner the population increases over time. See Figures 34, 35, and 36.

An easy way of interpreting the graphs is that the horizontal axis (x-axis) represents the starting population at a particular point in time, and the vertical axis (y-axis) the ending population at the end of each period (say one year) using the formula $x_{n+1} = rx_n(1 - x_n)$.

These graphs are based on the assumption that the population is in a restricted area that can accommodate a maximum population of 1.0. This is simply to illustrate the phenomenon; otherwise, one could achieve the same graphical results by working with a different restricted population number. You can think of it as an analysis of fish in a tank that can only accommodate a maximum of 1 million fish.

The other underlying assumptions are the ones typically applied by population ecologists, namely that the competition for food increases as the population grows. Therefore, at a particular point, the demand for food will outstrip the supply, resulting in starvation and death of some population members. As the population declines, the food supply increases again, resulting in renewed growth. The cycle continues repeating itself in this boom and bust fashion. But there are instances whereby the growth rate may be such that the births and deaths are equal, resulting in a state of equilibrium.

The parabolic shape shown in the graphs reflects the situation where the population grows, reaches maturity, and then starts declining and subsequently becomes extinct. The perpendicular line reflects a situation whereby the population grows steadily without any decline. The steps on the graph reflect the population changes that Feigenbaum was modeling.

Figure 34 shows steady growth in the population at a rate of 2.5, starting from a population of 0.2. The population converges at around 0.6000 and remains constant in perpetuity.

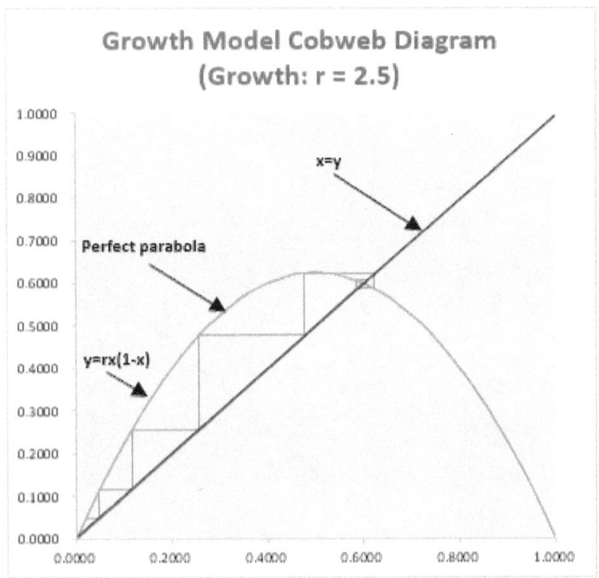

Figure 34: Simulation of a population using a growth rate of 2.5

Figure 35 shows a slightly different growth pattern using a growth rate of 3.3. Starting from a population of 0.2, the population settles at two different levels (the ones marked with black dots), namely, 0.4794 and 0.8236. The population switches between these two levels in perpetuity.

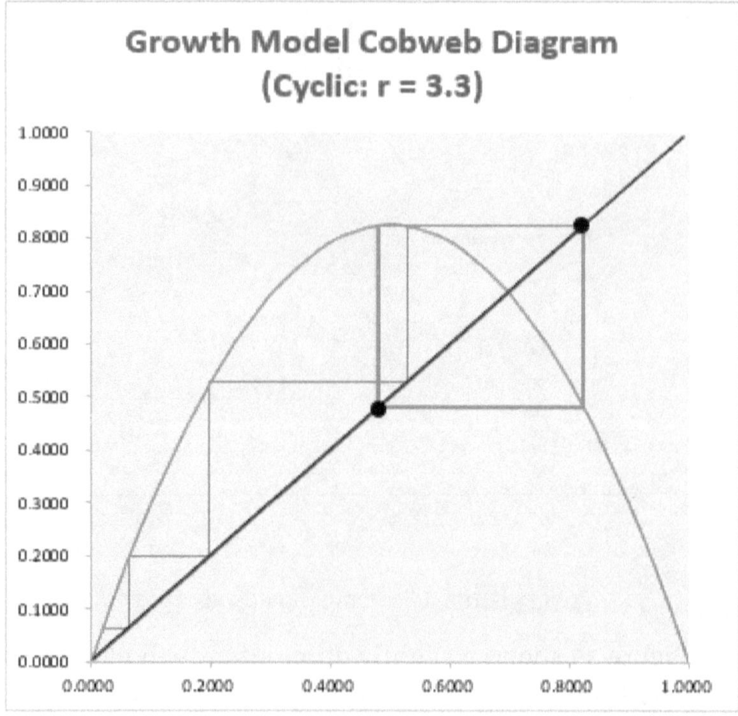

Figure 35: Simulation of a population using a growth rate of 3.3

And then something extraordinary happens when the growth rate reaches 4.0. The cycles become chaotic. There is no pattern whatsoever in the population after each cycle. It becomes impossible to predict the next population after each cycle. This chaotic behavior is reflected in Figure 36.

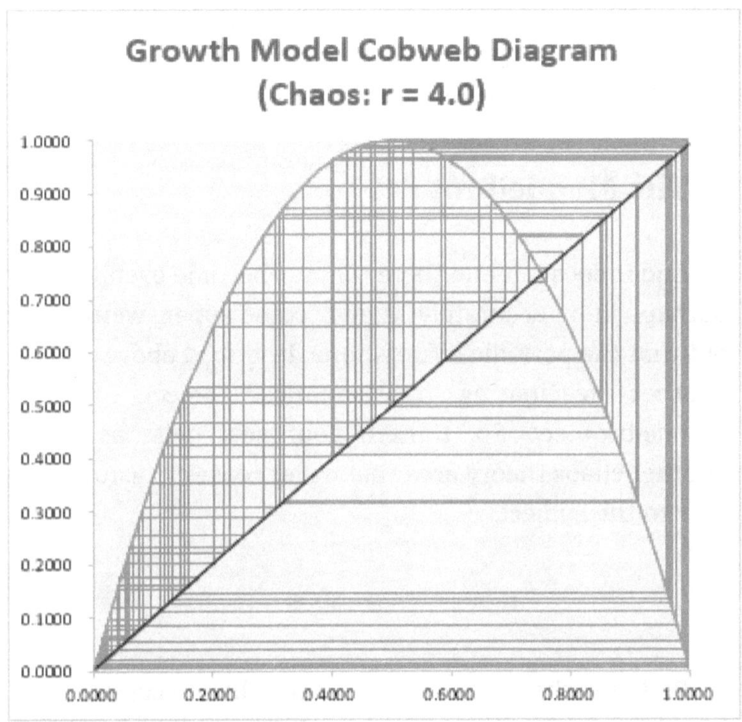

Figure 36: Simulation of a population using a growth rate of 4.0

Two exciting things emerge after multiple iterations of this scenario. The population aligns nicely with the left-hand side of the parabolic shape. Secondly, the populations seem to fall at the perpendicular line across the graph. This symmetry is quite astounding, considering the random nature of the population numbers at the end of every cycle.

It is noteworthy that these three graphs reflect the same bifurcations we observed earlier when we were discussing experiments by Lorenz (Figures 27, 28, and 29) and the logistic maps represented by Figures 30,31 and 32.

The Mandelbrot Set

And it does not end there. Things become even weirder, or perhaps more accurately, they become super-weird. It turns out that the periodic bifurcations described above are part of another fascinating mathematical object called the Mandelbrot set. So, tighten your seat belts as we enter another chaos theory area that never ceases to astonish those new to the subject.

* * *

Perhaps the best way to start the discussion on the Mandelbrot set is to look briefly at the gentleman's biography.

Benoit Mandelbrot was born in Warsaw, Poland, in 1924 to a Jewish family. His early life was challenging. He tried his best to escape the wrath of the Gestapo, who were after Jews.

He went to Ecole Normale in Paris, then transferred to the École Polytechnique and is renowned for his great aptitude in mathematics. He graduated from Polytechnique in 1947. He obtained a scholarship to study in the United States. In 1949 he graduated from the California Institute of Technology with a Master's degree in aeronautics. He returned to France to pursue doctoral studies. In 1952, he was awarded a doctorate in Mathematical Sciences by the University of Paris. He spent a few years in academia in France and the USA before joining IBM in 1958. He periodically taught at Harvard University while working at IBM

One of his assignments at IBM was to help the company resolve a perennial problem in data transmission over telephone lines. The transmission was never smooth. It was always affected by interference. The interference seemed quite random and was a big puzzle to scientists at IBM. However, thanks to newly invented computer technology, Benoit could run tests and generate voluminous data showing how the data transmission was happening. And because of the challenges of interpreting massive volumes of data, Benoit developed a way of printing the data on a graph.

When he started generating the printouts in the early stages, his colleagues thought the data was garbled. In some instances, they erased some of the data before handing him the computer printouts. And when Benoit discovered what was happening, he asked his colleagues not to interfere with the printouts.

When he closely examined the printouts, he saw a certain interesting pattern. The data points were scattered in a certain uniform shape. The data points distribution resembled what he had seen several years earlier in the so-called Cantor set.

The Cantor Set

The Cantor set is a straightforward mathematical set. It works like this. You take a line, divide it into three, then remove the middle section. Take the remaining two lines, divide each of them into three too, and discard the middle sections. If you repeat this exercise multiple times, you will end up with something that looks like Figure 37.

Figure 37: Cantor set

* * *

The resemblance of the disturbance signals to the Cantor set was quite intriguing to Benoit. He decided to go one step further.

Mandelbrot Fractal

Benoit Mandelbrot developed a formula close to the Cantor model for generating new numbers. The formula was as follows:

$$z_{n+1} = z_n^2 + C$$

Take a number. Multiply it by itself, and then add another constant number. Take the result that you get and repeat the process again. By doing this exercise repeatedly, say, for one million iterations, you would end up with a set of numbers that, when shown on a graph, would look like Figure 38.

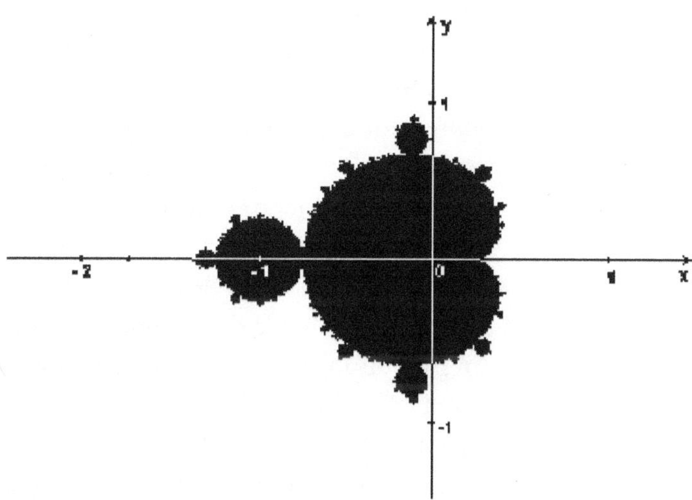

Figure 38: The Mandelbrot Set[53]

The one unique aspect of this graph (that we will not get into detail) is that it is drawn on a complex plain. In other words, the horizontal axis shows the real numbers that we are all used to, namely, 0,1,2,3,4, etc., moving to the right, and -1, -2, -3, etc. moving to the left. In other words, the normal numbers that usually appear on the x-axis of a graph. On the other hand, on the vertical axis are the complex numbers that we discussed in Chapter 11. In other words, the "i" numbers.

So, the numbers that Mandelbrot generated using his simple formula were both in respect of the real numbers and complex numbers.

And when he generated a printout of the millions of numbers churned out by the computer using the simple formula indicated above, the output was miraculous. The image looked like a bug. See Figure 38.

But what was most fascinating was that by looking closely at the bug-like image, Benoit saw that the image showed even smaller bugs of the same shape. And by zooming into the smaller bugs, they too contained even smaller bugs. The self-similarity at smaller and smaller scales was an amazing phenomenon. You can also experience the wonder by looking at one of the numerous images of the Mandelbrot set available on the internet.

There is a lot to say about the Mandelbrot set that would fill a whole book. But the purpose of our discussion here is to demonstrate the beauty that sometimes emerges from chaos.

* * *

Many things in nature are very similar to the bifurcations that mathematicians can generate using simple chaos formulae.

I hope you are as mesmerized as I am by these incredible images.

After I first came across these ideas, I could not help thinking very strongly that nature itself is indeed mathematical.

* * *

So, what does the Mandelbrot set have to do with the bifurcations that we looked at earlier? It turns out that the graph that we looked at is an extension of the Mandelbrot set. It occurs at the tip of the Mandelbrot set. This is not obvious because the Mandelbrot set is on the complex plane.

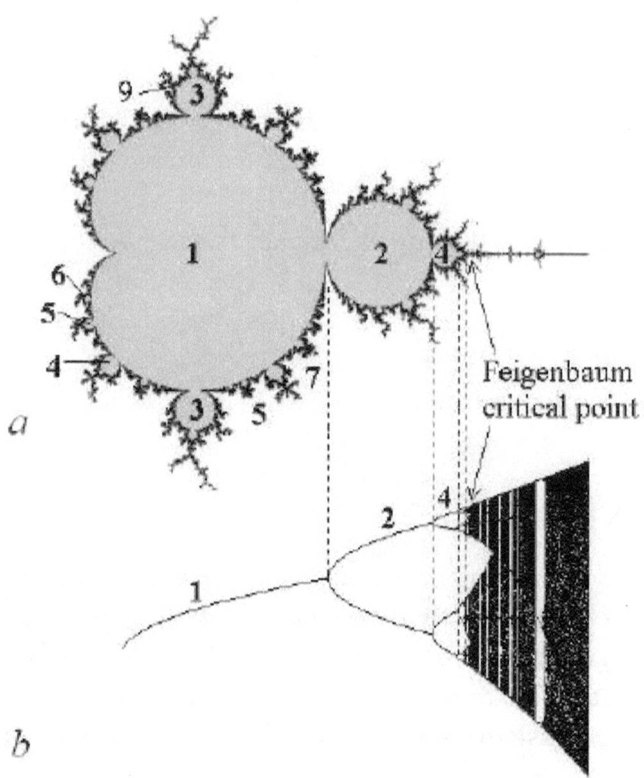

Figure 39: Mandelbrot Set and Logistic Map[54]

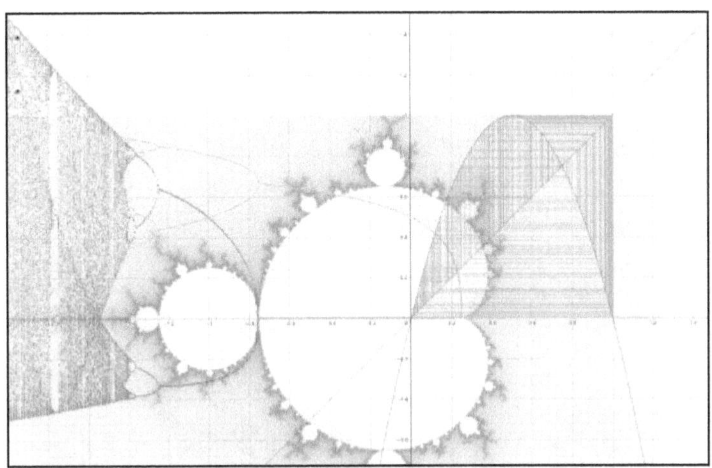

Figure 40: Matching of Mandelbrot Set and Bifurcations[55]

You can see the visual merging of the two images in the YouTube video developed by Eye of the Universe – Mandelbrot Fractal Zoom (e1091) (4k 60fps) at https://www.youtube.com/watch?v=pCpLWbHVNhk). This phenomenon is truly mind–boggling.

* * *

PART 3

What does it all mean?

CHAPTER 15

What does it all mean?

"Simplicity is a great virtue but it requires hard work to achieve it and education to appreciate it. And to make matters worse: complexity sells better."
—Edsger Wybe Dijkstra

I HOPE THAT READING THE PRECEDING science and mathematics basics has given you a different perspective of nature. And even more importantly, that your curiosity has been re-wakened to the point where you want to learn even more.

The amount of knowledge out there that is inaccessible to non-scientists and non-mathematicians is incredible. But with just a little effort, one can gain the essential basic understanding and wallow in the wonders of nature, some of which do not have a logical explanation. And even in mathematics, man is still far from exactitude. There are still many knowledge gaps waiting to be filled by humanity.

APPENDIX

Maths House of Horrors

"In real life, I assure you, there is no such thing as algebra. "
— Fran Lebowitz

Non-mathematicians are not allowed into the Maths House of Horrors unless accompanied by two of their grandparents, one of whom must be a psychiatrist or a certified pure mathematician. Any non-mathematician who trespasses into the house will be doing so at their own peril. The author will not be held accountable for any incidents of dizziness, shock, mental anguish, anxiety attacks, or any associated maladies that such trespassers may suffer.

* * *

Infinitesimal Calculi

These animals live in the domain of infinitely small numbers. Sometimes you see them; sometimes, you don't. It all depends on your state of mind unless you are a certified pure mathematician.

Here is the easiest way of thinking about infinitesimal calculi. You start by thinking big, then gradually think smaller and smaller until you cannot think anymore. That is the essence of infinitesimal calculi.

There are really two species of infinitesimal calculi. One is called "slope," and the other one "area." However, mathematicians prefer to call them "differentiation" and "integration," respectively. The species you will see in the Maths House of Horrors today is differentiation.

The beauty of these animals is that, despite their occasional disappearing act, they can tell you the precise length of curves (not those curves), but the mathematical kind that you see when you look at circles.

If you were to ask them to show you how to do it, this is what they would do (but please do not try it at home, unless in the presence of a pure mathematician).

Here, the basic idea is to demonstrate that a straightforward concept, such as what we discussed in Chapter 6, can get obfuscated beyond belief. If you followed the basic idea of determining the circumference of a circle, then what you see in this Maths House of Horrors is just a mirror image of what you already know.

Calculating the Circumference of a Circle[56]

Take a circle in an x, y plane as follows:

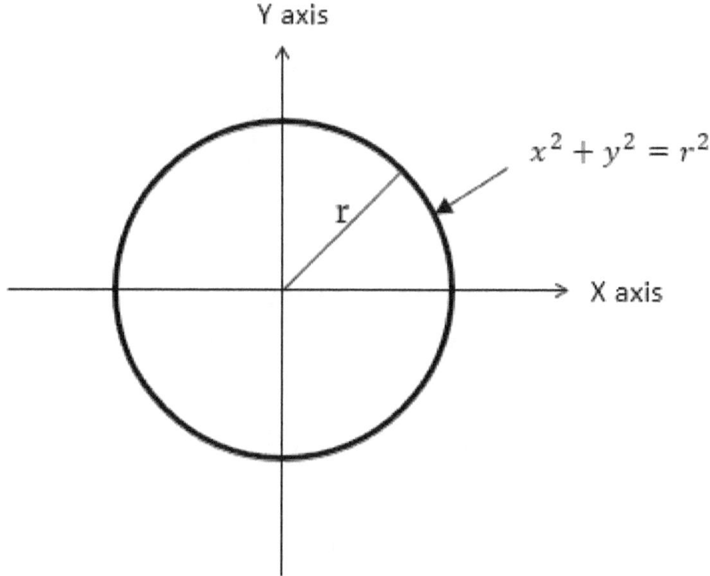

The equation $x^2 + y^2 = r^2$ can be re-written as follows:

$y^2 = r^2 - x^2$ or

$$y = \sqrt{r^2 - x^2}$$

Taking the derivative of the above equation with respect to x, we get:

$$\frac{dy}{dx} = \frac{-2x}{2\sqrt{r^2 - x^2}} = -\frac{x}{\sqrt{r^2 - x^2}}$$

Take one quadrant of the circle:

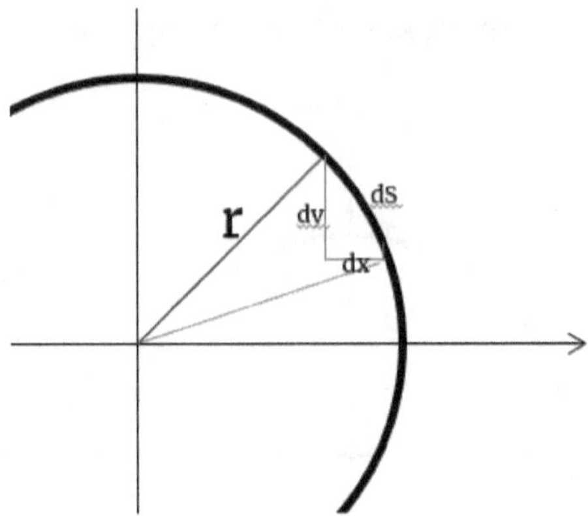

The length of a curve is given by the formula

$$dS = \sqrt{(dy)^2 + (dx)^2}$$

$$S = \int_0^r \sqrt{(dy)^2 + (dx)^2}$$

$$S = \int_0^r \sqrt{1 + (\frac{dy}{dx})^2} dx$$

Substitute $^{dy}/_{dx}$ to the above equation:

$$S = \int_0^r \sqrt{1 + (-\frac{x}{\sqrt{r^2 - x^2}})^2} dx$$

$$S = \int_0^r \sqrt{1 + \frac{x^2}{r^2 - x^2}}\, dx$$

$$S = \int_0^r \sqrt{\frac{r^2 - x^2 + x^2}{r^2 - x^2}}\, dx$$

$$S = \int_0^r \sqrt{\frac{r^2}{r^2 - x^2}}\, dx$$

$$S = r \int_0^r \frac{1}{\sqrt{r^2 - x^2}}\, dx$$

$$S = r[Sin^{-1}(\frac{x}{r})]_0^r$$

$$S = r[Sin^{-1}(\frac{r}{r}) - Sin^{-1}(\frac{0}{r})]$$

$$S = r[Sin^{-1}(1) - Sin^{-1}(0)]$$

$$S = r[\frac{\pi}{2} - 0]$$

$$S = \frac{1}{2}\pi r$$

Therefore, the circumference is:

$$C = 4S$$

$$C = 4(\frac{1}{2}\pi r)$$

$$C = 2\pi r$$

PI the Animal that Feeds on Humble Pie

The mathematical wizard, Leonhard Euler (1707-1783), determined that there was a relationship between *pi* and the reciprocals of the infinite series of squares of ordinal numbers, as shown below.[57]

$$\frac{\pi^2}{6} = \frac{1}{1^2} + \frac{1}{2^2} + \frac{1}{3^2} + \frac{1}{4} + \frac{1}{5^2} + \frac{1}{6^2} + \ldots$$

In 1995, two Canadian mathematicians, Peter Borwein and Simon Plouffe discovered the so-called Bailey–Borwein–Plouffe formula for *pi*:

$$\pi = \sum_{k=0}^{\infty} \left[\frac{1}{16^k} \left(\frac{4}{8k+1} - \frac{2}{8k+4} - \frac{1}{8k+5} - \frac{1}{8k+6} \right) \right]$$

If you do not get a headache after looking at this formula, count yourself as a very lucky mortal. It proves that you can maintain your sanity even after intense obfuscation.

And this is the point when you need to take a deep breath, then take a sip of your favorite beverage. Finally, congratulate yourself for going through the Maths House of Horrors and coming out of it mentally unscathed.

Meet Root Two the Beast

I don't know about you, but I find the name Root Two rather scary. Can you imagine meeting a beast with such a name in a dark alley in the back streets of the seedy side of Kirinyaga Road in Nairobi? I do not even want to think about it.

The good news is that according to Wildberger (2011), talking about the number $\sqrt{2}$ is a misnomer right from the get-go. The beast does not exist.

The truth is that there is no such number that, when squared, can give you 2. Indeed, even if you see a cage in the Maths House of Horrors with a sign saying "Root Two," you will not see anything if you look inside the cage. Root Two is a mathematical ghost of some sort.

$$* \quad * \quad *$$

Wildberger points out three ways of dealing with this problem: applied, algebraic, and analytic.[58]

The applied approach is simply saying that if $\sqrt{2}$ does not exist, then we should just use an approximation. And in fact, this would work superbly in several practical situations. For example, suppose you wanted a carpenter to build for you a stool measuring one foot by one foot (i.e., 1-foot square), and you wanted to decorate it with a gold-colored tape that runs diagonally across the middle of the stool. You could comfortably place an order for a tape measuring approximately 1.4 feet (which would be a good approximation of $\sqrt{2}$). Note that the square of 1.4 comes to 1.96, which is not very far from 2. If you were doing something that required a little more precision, then you could use a slightly better approximation of $\sqrt{2}$ such as 1.4142, which when squared comes to 1.99996164

The algebraic approach, on the other hand, is a bit thorny. It involves expanding our conception of algebra in which we create an object that plays the role of $\sqrt{2}$.

The analytic approach, introduced in mathematics by the Flemish mathematician, Simon Stevin in 1585, uses infinite decimals of $\sqrt{2}$. This would entail working with numbers such as the following:

1.41421356237309504880168872420969807785696718753 7694807317667973799...

And the reality is that there is no end to the number of decimals that we can get. Wildeberger (2011) says that this approach does not work properly. Such a large number does not make good sense, mainly because nobody has yet found a number that has the complete set of digits that give us $\sqrt{2}$. So, it should be really ignored, especially by ordinary mortals visiting the Maths House of Horrors for the first time.

I really wish Professor Wildberger was around during the days of Pythagoras. He could have prevented the *mathimatikoi* from murdering Hippasus. And if the *mathimatikoi* gang is still in existence, their leader ought to apologize to the world for the crime they committed in c–520BC.

BIBLIOGRAPHY

"Did This Man Cause Raila's Podium Collapse in Malindi?"
Nairobi News, March 28, 2016.
https://nairobinews.nation.co.ke/news/did-this-man.

"World Development Indicators." DataBank. The World Bank,
2020.
https://databank.worldbank.org/reports.aspx?source=2.

Boyer, Carl B. "Leonhard Euler." Encyclopædia Britannica.
Encyclopædia Britannica, inc., April 11, 2020.
https://www.britannica.com/biography/Leonhard-Euler.

CK-12 Foundation, -. "Heats of Vaporization and
Condensation." CK. CK-12 Foundation, April 30, 2014.
https://www.ck12.org/c/chemistry/heats-of-
vaporization-and-condensation/enrichment/Heat-of-
vaporization-Overview/.

Clegg, Brian. "The Dangerous Ratio." NRICH, 2004.
https://nrich.maths.org/2671.

Cognito. "Plant Hormones - Tropisms & Auxins
#77" *YouTube* video, 5:01. February 23, 2020.
https://www.youtube.com/watch?v=rKHIfsHX1aA

Complex Numbers. "15 - Complex Numbers & the Complex
Plane" *YouTube* video, 32:56. June 11, 2019.
https://www.youtube.com/watch?v=96C42d0yes8&t=153
2s

Coolman, Robert. "Euler's Identity: 'The Most Beautiful Equation.'"LiveScience. Purch, July 1, 2015. https://www.livescience.com/51399-eulers-identity.html.

Corn, Patrick, Hua Z Vee, and Christopher Williams. "Benford's Law." Brilliant Math & Science Wiki, 2020. https://brilliant.org/wiki/benfords-law/.

Darwin, Charles. *On the Origin of Species by Means of Natural Selection.* Edison, NJ, NJ: Chartwell Books, 2008.

Davies, Antony. "The Cobra Effect: Lessons in Unintended Consequences: Antony Davies, James R. Harrigan." FEE Freeman Article. Foundation for Economic Education, September 6, 2019. https://fee.org/articles/the-cobra-effect-lessons-in-unintended-consequences/.

Davies, Antony. "The Cobra Effect: Lessons in Unintended Consequences: Antony Davies, James R. Harrigan." FEE Freeman Article. Foundation for Economic Education, September 6, 2019. https://fee.org/articles/the-cobra-effect-lessons-in-unintended-consequences/.

Eliminate the Four Pests (1958), 2016. https://chineseposters.net/themes/four-pests.php.

Falk, Dan. "What Is Relativity? Einstein's Mind-Bending Theory Explained." NBCNews.com. NBCUniversal News Group, November 29, 2018. https://www.nbcnews.com/mach/science/what-relativity-einstein-s-mind-bending-theory-explained-ncna865496.

Fichter, Lynn, Steve Baedke, Eric Pyle, and Steve Whitmeyer. "James Madison University - Department of Geology & Environmental Science." Teaching Chaos/Complex Systems to Beginners, 2009. http://csmgeo.csm.jmu.edu/geollab/complexevolutionarysystems/BifurcationDiagram.htm.

Fleming, Amy. "The Secret Life of Plants: How They Memorise, Communicate, Problem Solve, and Socialise." The Guardian. Guardian News and Media, April 5, 2020. https://www.theguardian.com/environment/2020/apr/05

/smarty-plants-are-our-vegetable-cousins-more-intelligent-than-we-realise.

Foundation, CK-12. "12 Foundation." CK, 2020. https://www.ck12.org/book/ck-12-chemistry---basic/section/17.3/.

Frenay, Robert. *Pulse: How Nature Is Inspiring the Technology of the 21st Century Paperback*. London, UK: Little, Brown, 2006.

Frenay, Robert. *Pulse: How Nature Is Inspiring the Technology of the 21st Century Paperback*. London, UK: Little, Brown, 2006.

Gleick, James, and Ralph Losey. "What Chaos Theory Tell Us About e-Discovery and the Projected 'Information → Knowledge → Wisdom' Transition." e, May 20, 2016. https://e-discoveryteam.com/2016/05/20/what-chaos-theory-tell-us-about-e-discovery-and-the-projected-information-%E2%86%92-knowledge-%E2%86%92-wisdom-transition/.

Gleick, James. "Universality." Essay. In *Chaos: Making a New Science*, 160–60. New York, NY: Penguin Books, 1988.

Gleick, James. *Chaos: Making a New Science*. New York, NY: Penguin Books, 1988.

Gleick, James. *Chaos: Making a New Science*. New York, NY: Penguin Books, 1988.

Gleick, James. *Chaos: Making a New Science*. New York, NY: Penguin Books, 1988.

GSCE Physics. "Specific Latent Heat #28" *YouTube* video, 6:25. October 10, 2019. https://www.youtube.com/watch?v=3itqmCtmJPc

Hignett, Katherine. "The Devastation of Nagasaki and the Luck of Kokura: A Tale ..." Newsweek, 2018. https://www.newsweek.com/devastation-nagasaki-and-luck-kokura-tale-two-cities-1064991.

Hill, Theodore P. "A Statistical Derivation of the Significant-Digit Law." *Statistical Science* 10, no. 4 (1995): 354–63. https://doi.org/10.1214/ss/1177009869.

Hoffman, Paul. *The Man Who Loved Only Numbers: The Curious Story of Paul Erdos Andhis Search for Mathematical Truth*. Fourth Estate, London: Fourth Estate Limited, 1998.

Isaeva, O, and S Kuznetsov. "Complex Generalization of Approximate Renormalization ..." ResearchGate, 2004. https://www.researchgate.net/publication/228635212_Complex_generalization_of_approximate_renormalization_group_analysis_and_Mandelbrot_set_Thermodynamic_Analogy.

Kooser, Amanda. "Pi Dream: Google Employee Sets Record for Most Digits Calculated." CNET. CNET, March 14, 2019. https://www.cnet.com/news/pi-dream-googler-sets-record-for-most-digits-calculated/.

Kreston, Rebecca. "Paved With Good Intentions: Mao Tse-Tung's 'Four Pests' Disaster." Discover Magazine. Discover Magazine, October 15, 2019. https://www.discovermagazine.com/health/paved-with-good-intentions-mao-tse-tungs-four-pests-disaster.

Libretexts. "Fundamentals of Phase Transitions." Chemistry LibreTexts. Libretexts, July 14, 2020. https://chem.libretexts.org/Bookshelves/Physical_and_Theoretical_Chemistry_Textbook_Maps/Supplemental_Modules_(Physical_and_Theoretical_Chemistry)/Physical_Properties_of_Matter/States_of_Matter/Phase_Transitions/Fundamentals_of_Phase_Transitions.

Lowy, Franklin D. "Antimicrobial Resistance: the Example of Staphylococcus Aureus." *Journal of Clinical Investigation* 111, no. 9 (2003): 1265–73. https://doi.org/10.1172/jci18535.

Mancuso, Stefano, and Vanessa Di Stefano. *The Revolutionary Genius of Plants: a New Understanding of Plant Intelligence and Behavior*. New York, NY: Atria Books, 2018.

Maverick, J.B. "The 4 Countries That Produce the Most Chocolate." Investopedia. Investopedia, January 29, 2020. https://www.investopedia.com/articles/investing/093015/4-countries-produce-most-chocolate.asp.

Miller, Steven J. *Benford's Law: Theory and Applications*. Princeton, NJ: Princeton University Press, 2015.

Newcomb, Simon. "Note on the Frequency of Use of the Different Digits in Natural Numbers." *American Journal of*

Mathematics 4, no. 1/4, (1881): 39.
https://doi.org/10.2307/2369148.

Pantsov, Alexander, and Steven I. Levine. *Mao: The Real Story.* New York, NY: Simon & Schuster, 2013.

Parrish, Shane. "The Butterfly Effect: Everything You Need to Know About This Powerful Mental Model." Farnam Street, November 27, 2019. https://fs.blog/2017/08/the-butterfly-effect/.

Parrish, Shane. "The Butterfly Effect: Everything You Need to Know About This Powerful Mental Model." Farnam Street, November 27, 2019. https://fs.blog/2017/08/the-butterfly-effect/.

Pratchett, Terry, and Neil Gaiman. *Good Omens: The Nice and Accurate Prophecies of Agnes Nutter, Witch.* London, UK: Corgi books, 2019.

Principles, Math. "Circle - Circumference Derivation." Math Principles, January 1, 1970. http://www.math-principles.com/2013/02/circle-circumference-derivation.html.

Sandron, Frédéric. "Do Populations Conform to the Law of Anomalous Numbers ?" CORE, January 1, 1970. https://core.ac.uk/display/81233726.

Sapolsky, Robert. "21. Chaos and Reductionism." *Stanford.* Lecture, 2020. https://www.youtube.com/watch?v=_njf8jwEGRo&list=PLD7E21BF91F3F9683&index=21.

Singleton, Tommie. "Understanding and Applying Benfords Law." ISACA Journal Archives, 2011. https://www.isaca.org/resources/isaca-journal/past-issues/2011/understanding-and-applying-benfords-law.

Sparks, Ben. "Bifurcation Diagram (Logistic Map)." GeoGebra, 2020. https://www.geogebra.org/m/wQbHRgye.

Sprakel, Niek. "Mandelbrot-Bifurcation." GeoGebra, 2020. https://www.geogebra.org/m/m8kd5bvb.

Wildberger, Norman J. "Pythagoras' Theorem (a) | Math History | NJ Wildberger." *Insights into Mathematics.* Lecture, 2020. https://www.youtube.com/watch?v=dW8Cy6WrO94.

Wildberger, Norman J. "Pythagoras' Theorem (a) | Math
    History | NJ Wildberger." *Insights into Mathematics.*
    Lecture, 2020.
    https://www.youtube.com/watch?v=dW8Cy6WrO94.

NOTES

1 "Did This Man Cause Raila's Podium Collapse in Malindi?"
    Nairobi News, March 28, 2016.
    https://nairobinews.nation.co.ke/news/did-this-man.
2 Falk, Dan. "What Is Relativity? Einstein's Mind-Bending
    Theory Explained." NBCNews.com. NBCUniversal News
    Group, November 29, 2018.
    https://www.nbcnews.com/mach/science/what-
    relativity-einstein-s-mind-bending-theory-explained-
    ncna865496.
3 Cognito. "Plant Hormones - Tropisms & Auxins
    #77" *YouTube* video, 5:01. February 23, 2020.
    https://www.youtube.com/watch?v=rKHIfsHX1aA
4 Mancuso, Stefano, and Vanessa Di Stefano. *The Revolutionary
    Genius of Plants: a New Understanding of Plant Intelligence
    and Behavior.* New York, NY: Atria Books, 2018.
5 Fleming, Amy. "The Secret Life of Plants: How They
    Memorise, Communicate, Problem Solve and Socialise."
    The Guardian. Guardian News and Media, April 5, 2020.
    https://www.theguardian.com/environment/2020/apr/05
    /smarty-plants-are-our-vegetable-cousins-more-
    intelligent-than-we-realise.
6 Libretexts. "Fundamentals of Phase Transitions." Chemistry
    LibreTexts. Libretexts, July 14, 2020.

https://chem.libretexts.org/Bookshelves/Physical_and_
Theoretical_Chemistry_Textbook_Maps/Supplemental_
Modules_(Physical_and_Theoretical_Chemistry)/Physi
cal_Properties_of_Matter/States_of_Matter/Phase_Tra
nsitions/Fundamentals_of_Phase_Transitions.

7 Foundation, CK-12. "12 Foundation." CK, 2020.
https://www.ck12.org/book/ck-12-chemistry---
basic/section/17.3/.

8 CK-12 Foundation, -. "Heats of Vaporization and
Condensation." CK. CK-12 Foundation, April 30, 2014.
https://www.ck12.org/c/chemistry/heats-of-
vaporization-and-condensation/enrichment/Heat-of-
vaporization-Overview/.

9 GSCE Physics. "Specific Latent Heat #28" *YouTube* video, 6:25.
October 10, 2019.
https://www.youtube.com/watch?v=3itqmCtmJPc

10 Gleick, James. "Universality." Essay. In *Chaos: Making a New
Science*, 160–60. New York, NY: Penguin Books, 1988.

11 Kreston, Rebecca. "Paved With Good Intentions: Mao Tse-
Tung's 'Four Pests' Disaster." Discover Magazine.
Discover Magazine, October 15, 2019.
https://www.discovermagazine.com/health/paved-with-
good-intentions-mao-tse-tungs-four-pests-disaster.

12 Ibid

13 Eliminate the Four Pests (1958), 2016.
https://chineseposters.net/themes/four-pests.php.

14 Ibid

15 Ibid

16 Pantsov, Alexander, and Steven I. Levine. *Mao: The Real Story*.
New York, NY: Simon & Schuster, 2013.

17 Davies, Antony. "The Cobra Effect: Lessons in Unintended
Consequences: Antony Davies, James R. Harrigan." FEE
Freeman Article. Foundation for Economic Education,
September 6, 2019. https://fee.org/articles/the-cobra-
effect-lessons-in-unintended-consequences/.

18 Ibid

19 Maverick, J.B. "The 4 Countries That Produce the Most
Chocolate." Investopedia. Investopedia, January 29,

20 2020.
    https://www.investopedia.com/articles/investing/093
    015/4-countries-produce-most-chocolate.asp

[20] Ibid

[21] Darwin, Charles. *On the Origin of Species by Means of Natural Selection*. Edison, NJ, NJ: Chartwell Books, 2008.

[22] Frenay, Robert. *Pulse: How Nature Is Inspiring the Technology of the 21st Century Paperback*. Lndon, UK: Little, Brown, 2006.

[23] "World Development Indicators." DataBank. The World Bank, 2020.
    https://databank.worldbank.org/reports.aspx?source=2.

[24] Frenay, Robert. *Pulse: How Nature Is Inspiring the Technology of the 21st Century Paperback*. London, UK: Little, Brown, 2006.

[25] Lowy, Franklin D. "Antimicrobial Resistance: the Example of Staphylococcus Aureus." *Journal of Clinical Investigation* 111, no. 9 (2003): 1265–73.
    https://doi.org/10.1172/jci18535.

[26] Singleton, Tommie. "Understanding and Applying Benfords Law." ISACA Journal Archives, 2011.
    https://www.isaca.org/resources/isaca-journal/past-issues/2011/understanding-and-applying-benfords-law.

[27] Hill, Theodore P. "A Statistical Derivation of the Significant-Digit Law." *Statistical Science* 10, no. 4 (1995): 354–63.
    https://doi.org/10.1214/ss/1177009869.

[28] Newcomb, Simon. "Note on the Frequency of Use of the Different Digits in Natural Numbers." *American Journal of Mathematics* 4, no. 1/4 (1881): 39.
    https://doi.org/10.2307/2369148.

[29] Sandron, Frédéric. "Do Populations Conform to the Law of Anomalous Numbers?" CORE, January 1, 1970.
    https://core.ac.uk/display/81233726.

[30] Miller, Steven J. *Benford's Law: Theory and Applications*. Princeton, NJ: Princeton University Press, 2015.

[31] Corn, Patrick, Hua Z Vee, and Christopher Williams. "Benford's Law." Brilliant Math & Science Wiki, 2020.
    https://brilliant.org/wiki/benfords-law/.

[32] Kooser, Amanda. "Pi Dream: Google Employee Sets Record for Most Digits Calculated." CNET. CNET, March 14, 2019. https://www.cnet.com/news/pi-dream-googler-sets-record-for-most-digits-calculated/.

[33] Wildberger, Norman J. "Pythagoras' Theorem (a) | Math History | NJ Wildberger." *Insights into Mathematics.* Lecture, 2020. https://www.youtube.com/watch?v=dW8Cy6WrO94.

[34] Clegg, Brian. "The Dangerous Ratio." NRICH, 2004. https://nrich.maths.org/2671.

[35] Complex Numbers. "15 - Complex Numbers & the Complex Plane" *YouTube* video, 32:56. June 11, 2019. https://www.youtube.com/watch?v=96C42d0yes8&t=1532s

[36] Coolman, Robert. "Euler's Identity: 'The Most Beautiful Equation'." LiveScience. Purch, July 1, 2015. https://www.livescience.com/51399-eulers-identity.html.

[37] Boyer, Carl B. "Leonhard Euler." Encyclopædia Britannica. Encyclopædia Britannica, inc., April 11, 2020. https://www.britannica.com/biography/Leonhard-Euler.

[38] Ibid

[39] Ibid

[40] Ibid

[41] Pratchett, Terry, and Neil Gaiman. *Good Omens: The Nice and Accurate Prophecies of Agnes Nutter, Witch.* London, UK: Corgi books, 2019.

[42] Parrish, Shane. "The Butterfly Effect: Everything You Need to Know About This Powerful Mental Model." Farnam Street, November 27, 2019. https://fs.blog/2017/08/the-butterfly-effect/.

[43] Gleick, James. *Chaos: Making a New Science.* New York, NY: Penguin Books, 1988.

[44] Hignett, Katherine. "The Devastation of Nagasaki and the Luck of Kokura: A Tale ..." Newsweek, 2018. https://www.newsweek.com/devastation-nagasaki-and-luck-kokura-tale-two-cities-1064991.

[45] Parrish, Shane. "The Butterfly Effect: Everything You Need to Know About This Powerful Mental Model." Farnam Street, November 27, 2019. https://fs.blog/2017/08/the-butterfly-effect/.

[46] Gleick, James. *Chaos: Making a New Science*. New York, NY: Penguin Books, 1988.

[47] Sapolsky, Robert. "21. Chaos and Reductionism." *Stanford*. Lecture, 2020. https://www.youtube.com/watch?v=_njf8jwEGRo&list=PLD7E21BF91F3F9683&index=21.

[48] Gleick, James. *Chaos: Making a New Science*. New York, NY: Penguin Books, 1988.

[49] Ibid

[50] Sparks, Ben. "Bifurcation Diagram (Logistic Map)." GeoGebra, 2020. https://www.geogebra.org/m/wQbHRgye.

[51] Ibid

[52] Fichter, Lynn, Steve Baedke , Eric Pyle, and Steve Whitmeyer. "James Madison University - Department of Geology & Environmental Science." Teaching Chaos/Complex Systems to Beginners, 2009. http://csmgeo.csm.jmu.edu/geollab/complexevolutionarysystems/BifurcationDiagram.htm.

[53] Gleick, James, and Ralph Losey. "What Chaos Theory Tell Us About e-Discovery and the Projected 'Information → Knowledge → Wisdom' Transition." e, May 20, 2016. https://e-discoveryteam.com/2016/05/20/what-chaos-theory-tell-us-about-e-discovery-and-the-projected-information-%E2%86%92-knowledge-%E2%86%92-wisdom-transition/.

[54] Isaeva, O, and S Kuznetsov. "Complex Generalization of Approximate Renormalization ..." ResearchGate, 2004. https://www.researchgate.net/publication/228635212_Complex_generalization_of_approximate_renormalization_group_analysis_and_Mandelbrot_set_Thermodynamic_Analogy.

[55] Sprakel, Niek. "Mandelbrot-Bifurcation." GeoGebra, 2020. https://www.geogebra.org/m/m8kd5bvb.

---

56 Principles, Math. "Circle – Circumference Derivation." Math Principles, January 1, 1970. http://www.math-principles.com/2013/02/circle-circumference-derivation.html.

57 Hoffman, Paul. *The Man Who Loved Only Numbers: The Curious Story of Paul Erdos Andhis Search for Mathematical Truth.* Fourth Estate, London: Fourth Estate Limited, 1998.

58 Wildberger, Norman J. "Pythagoras' Theorem (a) | Math History | NJ Wildberger." *Insights into Mathematics.* Lecture, 2020. https://www.youtube.com/watch?v=dW8Cy6WrO94.

BOOKS BY THIS AUTHOR

Shamba Shenanigans: A Collection Of Riveting True Stories

A collection of riveting true-life experiences. Some of the stories are hilarious; others are thought-provoking, while others are likely to evoke different emotions as the story unfolds. Each story has one or more useful life lessons.

The Endless Search For More: A Collection Of True Stories On Money Matters

A collection of true stories that revolve around our continuous search for "more." And while this trait is essential for the long-term sustainability of humanity, John Mucai suggests that we must always strive to appropriately calibrate our desires. And more importantly, adopt a problem-solving mindset in our neverending quest for "more."

Seeking The Right Path: A Search For Spiritual Enlightenment

This book is a chronicle of a personal search for spiritual enlightenment. John Mucai starts by finding out what religion means. He then looks at the different religions and zeros on five major ones: Christianity, Islam, Hinduism, and Buddhism. These religions have a combined following comprising about 80% of the world's population. He looks at their beliefs, practices, and sacred texts.

And most importantly, the many complex questions that emerge from the texts. While the book would be of immense interest to theologians, it is not a book on theology. Instead, it is an attempt by the author to seek spiritual enlightenment by sifting through the religious literature freely available to any ordinary citizen of the world. The author's findings are illuminating and, hopefully, give believers and non-believers a new perspective on religion.

Historical Snapshots Of The Great: What Can We Learn From Them?

The quality of life that we enjoy today is a function of the many commendable actions done by individuals in different spheres of life. Some of these people came before us many years ago, while others live amongst us. This book explores the lives of some significant historical figures to find out whether they had any common attributes that we can emulate.

Multiple Dilemmas: A Fictional Story Of Multiple Ethical Dilemmas Based On True Historical Events

Multiple Dilemmas is a thriller based on historical events that raise significant ethical questions. The book delves deeply into challenging situations where ethical considerations are paramount, but the right choices are unclear. The twists and turns in the story will keep the reader entranced for several hours.

Ngurario: A Traditional Kikuyu Marriage Experience

Ngurario is a true story of the multiple steps that John and Susan went through to formalize their marriage according to Kikuyu traditions. The book delves deeply into the drama, excitement, and joy they experienced along the way, right up to the final step in the journey, namely, an elaborate and colorful ceremony called ngurario.

One Day In The Year 3000

Nobody knows what the future holds one thousand years from now. But one can make some wild guesses. This book peeks into that distant future.

Stratagem: Developing A Strategic Mindset

Have you ever attended a strategy meeting and wondered whether everyone in the forum had a clear understanding of what strategy meant? If you have, you are not alone. Interestingly, many such meetings roll on smoothly with impressive outcomes. That phenomenon is the ninth wonder of the world. Some participants probably spend many hours after the meeting engrossed in self-doubt or guilt, depending upon how loudly they spoke during the session. A cold or hot beverage usually works wonders during such moments of self-reflection. If you are one of those who experience self-doubt but usually emerge from strategy discussions with your conscience intact, you must count your blessings. You are a brave survivor. But whatever category you belong to, you have the cure for strategy-fuzziness right on your fingertips. Stratagem describes strategy with exceptional lucidity. Well-thought-out strategies are not only essential for business success; they are critical for success in personal life.

Archetypes Of Human Existence: A New Perspective

Out of the more than seven billion people who inhabit the earth, no two are exactly the same. Even tweens have differences. Every individual has been bestowed by nature's unique attributes. And yet, the behavior of human beings can be reduced to a few archetypes. At the heart of the matter, each human being is one single entity comprised of a mind and a physical body, a mind that yearns for happiness, and a body that longs for sustenance. And it is the interplay of these two needs that creates the different archetypes of humans. This book explores a few of the archetypes. It discusses how the ideas around archetypes converge to offer a new perspective on fundamental questions that existentialists have been grappling with for ages. The people described in the second chapter of this book are entirely fictitious. Any resemblance of their names to real people is purely coincidental. However, the characters are real and live among us. You may recognize some of them in your local community, your network of friends, or even in other human networks to which you are directly or indirectly connected.

Number One: Nothing Else Seems To Count

In the modern, highly competitive world, doing well in any competition is not enough. Being number one is what counts. This book traces the lives of five colorful individuals.

They are winners in their unique ways from the early stages of their lives. We experience, intimately, the twists and turns that occur as they enter early adulthood and get embroiled in a contest anchored in the pursuit of business success and love. At some point, each character will realize that things can become highly complex, emotionally draining, and even dangerous when love is in the mix. The outcome of their respective pursuits to be "number one" is astounding. Indeed, the way the story ends offers readers tremendous food for thought.

Fun and Grit: Encounters of Farming Hobbyists

The stories in this book are primarily about people–the people of the shamba (small farm). One of my insights after working for one of the biggest multinational companies and dabbling in a small–scale farming hobby is that every experience we go through in life, whether pleasant or unpleasant, is what gives life its flavor.

Indeed, some of the unpleasant experiences add more spice to life. Having a nice laugh about something is the magic trick in many cases. Laughter is undoubtedly the best medicine for the soul.

INDEX